CULTURE AND COSMOS
http://www.CultureAndCosmos.org

Culture and Cosmos is published twice a year, in northern spring/summer and autumn/winter, in association with the Sophia Centre for the Study of Cosmology in Culture, University of Wales Trinity Saint David.

Contributions and editorial correspondence should be addressed to:
editors@cultureandcosmos.org

Editor: Dr. Nicholas Campion, the Editor of *Culture and Cosmos*, Faculty of the Humanities and the Performing Arts, University of Wales Trinity Saint David, Lampeter, Ceredigion, Wales, SA48 7ED, UK.
E Mail **n.campion@uwtsd.ac.uk**

Deputy Editor: Dr. Jennifer Zahrt
Editorial Board: Dr. Silke Ackermann, Professor Anthony F. Aveni, Dr. Giuseppe Bezza, Dr. David Brown, Professor Charles Burnett, Dr. Hilary M. Carey, Dr. John Carlson, Dr Patrick Curry Professor Robert Ellwood, Dr. Germana Ernst, Dr. Ann Geneva, Professor Joscelyn Godwin, Dr. Dorian Greenbaum, Dr. Jacques Halbronn, Robert Hand, Dr Jarita Holbrook, Professor Michael Hunter, Professor Ronald Hutton, Dr Peter Kingsley, Dr. Edwin C. Krupp, Dr. J. Lee Lehman, Dr. Lester Ness, Professor P. M. Rattansi, Professor James Santucci, Robert Schmidt, Dr. Fabio Silva, Dr. Lorenzo Smerillo, Professor Richard Tarnas, Dr. Graeme Tobyn, Dr. David Ulansey, Robin Waterfield, Dr. Charles Webster, Dr. Graziella Federici Vescovini, Dr. Angela Voss, Dr. Paola Zambelli, Robert Zoller.
Technical assistance: Dr. Frances Clynes
Subscriptions:
For two issues: Individuals £20*
Institutions £40
http://www.cultureandcosmos.org/subscription.html

Payment for hard copy is online via PayPal. For bank transfer apply to the editor.
*Members of the British Astronomical Association, The Astrological Association and The Historical Association are entitled to a discount. Please enquire.

Contributors Guidelines: Please see http://www.cultureandcosmos.org/submissions.html
.

Front cover: Claude Lévi-Strauss: see p. 143

Published by Culture and Cosmos, Faculty of the Humanities and the Performing Arts, University of Wales Trinity Saint David, Lampeter, Ceredigion, Wales, SA48 7ED, UK.
© **Culture and Cosmos 2016**
Printed by Lightning Source.

The Sophia Centre
http://www.uwtsd.ac.uk/sophia/

The Sophia Centre for the Study of Cosmology in Culture is an academic centre within the Faculty of Humanities and the Performing Arts at the University of Wales Trinity Saint David.

The Centre's academic goals are

- 'to pursue research, scholarship and teaching in the relationship between astrological, astronomical and cosmological beliefs and theories, and society, politics, religion and the arts, past and present' and
- 'to undertake the academic and critical examination of astrology and its practice'.

The Centre's wider goal is stated in its title – to 'study cosmology in culture'. In a traditional sense, a cosmology is a world view, an understanding of the cosmos which informs individual and social action and ideology. The Centre promotes research in the subject area, holds seminars and conferences, publishes scholarly material, is associated with Sophia Centre Press and supervises PhD students.

The Centre's teaching is focused on the MA Cultural Astronomy and Astrology. For further information see
http://www.uwtsd.ac.uk/ma-cultural-astronomy-astrology/

CULTURE AND COSMOS

www.CultureAndCosmos.org

Editor Nicholas Campion
Vol. 18 No. 2 Autumn/Winter 2014   ISSN 1368-6534

Published in Association with
The Sophia Centre for the Study of Culture in Cosmology,
Faculty of Humanities and the Performing Arts
University of Wales Trinity Saint David
http://www.uwtsd.ac.uk/sophia/

Editorial

This issue of *Culture and Cosmos* includes seven significant papers on the history of astrology, covering a range of periods and approaches. Roger Beck's 'The Ancient Mithraeum as a Model Universe. Part 2', touches on archaeoastronomy and classical religion.[1] Helena Avelar and Charles Burnett's analysis of a twelfth century horoscope cast by Abraham the Jew examines the technical practice of medieval astrology. Lindsay Starkey's paper on Mellin de Saint-Gelais and John Calvin, and Scott Hendrix's on Galileo, concern theoretical contexts for the European astrology of the middle ages and Renaissance. Richard Angelo Bergen's *'Paradise Lost* and the Descent of Urania: from Astrology to Allegory' deals with literature, Hakan Kirkoğlu's *'Ilm-i nudjum* and 18th century Ottoman Court Politics' examines the political uses of astrology, and Graham Douglas's 'Trystes Cosmologiques: When Lévi-Strauss Met the Astrologers' explores one of the twentieth century's most important anthropologist's attitudes to astrology.[2] Anything Lévi-Strauss said about astrology is of interest by definition, in view of his authorship of a remarkable series of seminal works (*The Elementary Structures of Kinship* (1949), *Tristes tropiques* (1955), *Structural Anthropology* (1961), *Mythologiques* (1964), *The Raw and the Cooked* (1964), and *The Savage Mind* (1966). Lévi-

---

[1] Part 1 was published as Roger Beck, 'The Ancient Mithraeum as a Model Universe. Part 1', in *Heavenly Discourses,* ed. Nicholas Campion (Lampeter: Sophia Centre Press, 2016), pp. 21–31.

[2] For comparison see Nicholas Campion, 'Surrealist Cosmology: André Breton and Astrology', *Culture and Cosmos* 6, no. 2 (Autumn/Winter 2002): pp. 45–56.

Strauss, coming last chronologically in this journal, also has the last word. In response to a question about the surrealist André Breton, Lévi-Strauss replied:

> I knew André Bréton well – we were very close for a period of time; but I won't go as far as him. I wouldn't say that it holds [secrets] – but it is perhaps one of the signs that secrets exist which we don't understand, and I feel impelled to say, that we will doubtless never understand.

It is precisely this lack of understanding which motivates historians: with just a little more evidence, we hope, perhaps we will understand the world a little better. And perhaps, then, the papers in this issue will take us a little closer to understanding astrology's appeal, claims, role, nature, function, ideology, world-view and cultural significance.

Dr Nicholas Campion,
University of Wales Trinity Saint David

The Ancient Mithraeum as a Model Universe
Part 2

Roger Beck

Abstract: The aim of this two-part article is to confirm Porphyry's claim, in his essay *On the Cave of the Nymphs in the Odyssey'*, concerning the design and function of the ancient mithraeum: that the Mithraists 'perfect their initiate by inducting him into a mystery of the descent of souls and their exit back out again, calling the place a "cave"'; further, that the mithraeum/cave fulfilled that function by being an 'image of the cosmos', in which 'the things contained, by their proportionate arrangement', served as 'symbols of the elements and climates of the cosmos'.[1] Rather than attempting to recapitulate Part 1 (published in the Proceedings of the Heavenly Discourses Conference, Bristol 2010), I shall resume the argument directly.[2] Part 2 discusses in particular the significance of the solstitial diameter linking Cancer to Capricorn across the mithraeum's central aisle; the images of the torchbearers, Cautes (raised torch = ascent) and Cautopates (lowered torch = descent) and the mid-bench niches on either side of the aisle; the disposition of the images of the planets in the Mithraea of the Seven Spheres and the Seven Gates at Ostia; and finally the intent of the arcades of niches beneath the unusually elevated benches in the mithraeum at Vulci in Etruria.

We should now turn our attention to the second of the two key diameters of the cosmos (see Fig. 1). The first diameter, already discussed, is the equinoctial diameter, joining the spring equinox at the start of Aries to the autumn equinox at the start of Libra. In the diagram it appears as a thick black line.

The other diameter is the solstitial diameter, joining the summer solstice at the start of Cancer to the winter solstice at the start of Capricorn. It too is shown as a thick black line.

---

[1] Porphyry, *On the Cave of the Nymphs in the Odyssey* 6, ed. and trans. Seminar Classics 609 SUNY Buffalo (Buffalo: Arethusa, 1969).

[2] Roger Beck, 'The Ancient Mithraeum as Model Universe, Part 1', in *Heavenly Discourses*, ed. Nicholas Campion (Lampeter: Sophia Centre Press, 2016), pp. 21-31 [hereafter MMU Part 1].

Roger Beck 'The Ancient Mithraeum as a Model Universe Part 2', *Culture and Cosmos*, Vol. 18, no. 2, Spring/Summer 2014, pp. 3-18.
www.CultureAndCosmos.org

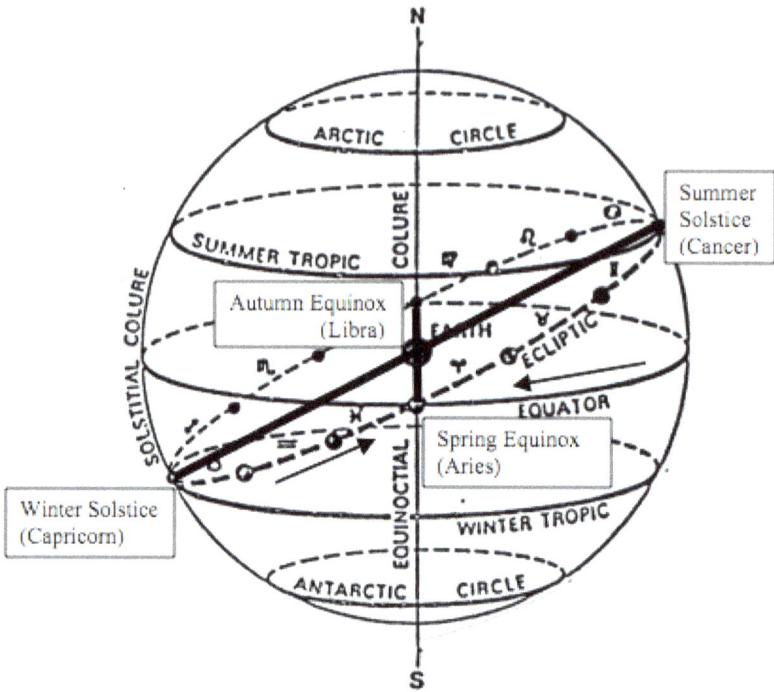

Fig. 1. Celestial sphere as conceived in basic Greek astronomy; author's drawing.

In the microcosm of the mithraeum the equinoctial diameter is the central aisle running between the two side-benches (see Fig. 2). It links the tauroctonous Mithras in the cult-niche to the mithraeum's entrance.

How is the solstitial diameter realized in the mithraeum? By the logic of geometry – or applied uranometry, if you prefer – it must be a line at right angles to the median line of the aisle, running from the mid point of the 'northern' bench (i.e. the bench representing the semicircle of the ecliptic/zodiac to the north of the ecliptic) to the mid point of the 'southern' bench (i.e. the bench representing the semicircle to the south). By the same logic, these mid points represent – and by representation become – the solstices: the summer solstice in Cancer and the winter solstice in Capricorn. Are the solstices marked in any way, other than by the solstitial signs, Cancer and Capricorn? In several mithraea, including

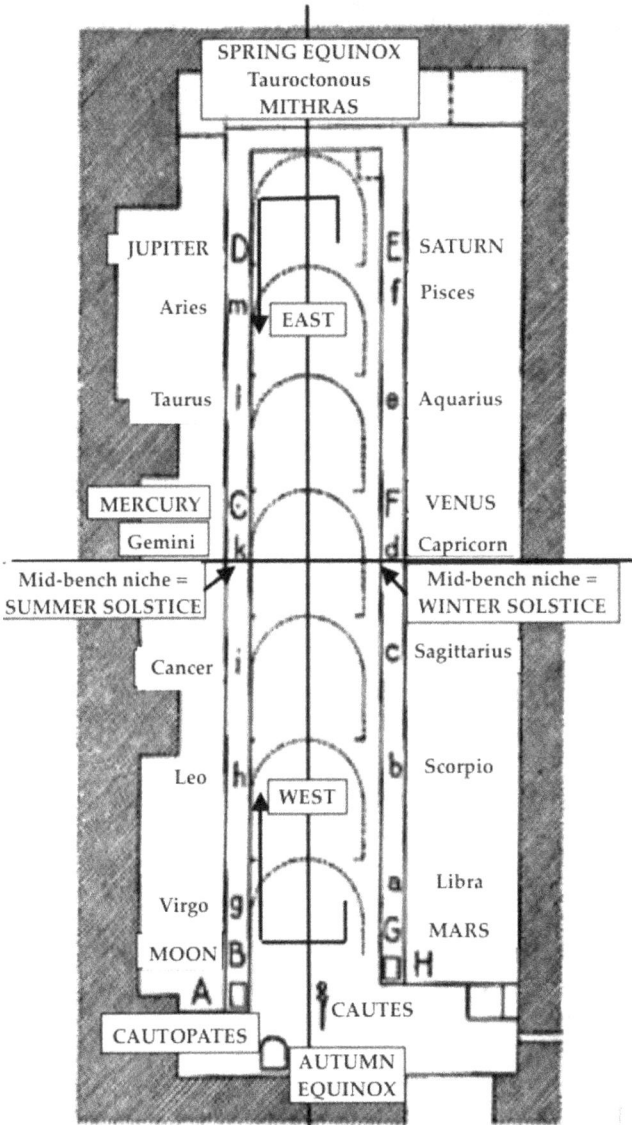

Fig. 2. Plan of the Mithraeum of the Seven Spheres (Sette Sfere), Ostia.[3]

---

[3] Author's drawing, superimposed on a photocopy of Fig. 71, p. 122 of Vol. 1 of M. J. Vermaseren, *Corpus Inscriptionum et Monumentorum Religionis Mithriacae* (The Hague: Martinus Nijhof, 1956).

the Mithraeum of the Seven Spheres (Sette Sfere) in Ostia (Fig. 2), we find in the two benches, and so facing each other at approximately the mid point, a small niche.[4] Do these opposed niches serve symbolically as the solstices, the summer solstice in Cancer on the northern bench to Mithras's right and the winter solstice in Capricorn on the bench to Mithras's left? Let us suppose provisionally that they do. They are in the correct positions, and they serve no obvious functional or structural purpose there.

To advance the argument, we must re-introduce our second passage from Porphyry's essay, 'On the Cave of the Nymphs in the *Odyssey*', quoting it here at somewhat greater length:

> To Mithras as his proper seat (*oikeian kathedran*) they assigned the equinoxes.... As creator and master of genesis, Mithras is set on the equator with the northern signs to his right and the southern signs to his left.... They set Cautes to the south because of its heat and Cautopates to the north because of the coldness of its wind.[5]

Who, or what, are Cautes and Cautopates, and why is the former 'set... to the south because of its heat' and the latter 'to the north because of the coldness of its wind'? The first question is easily answered. They are minor deities who accompany Mithras.[6] They are identical in appearance and pose, except that one, Cautes, carries a raised torch, the other,

---

[4] New on-line resources are now available. See in particular Eric Taylor's website of the Ostian mithraea:
http://www.ostia-antica.org/dict/topics/mithraea/mithraea.htm [accessed 21 March 2016].
For the Ostian Mithraeum of the Seven Spheres (Sette Sfere) see:
http://www.ostia-antica.org/regio2/8/8-6.htm [accessed 21 March 2016].
See also Roger Pearse's site of all Mithraic monuments:
http://roger-pearse.com/mithras/display.php?page=selected_monuments [accessed 21 March 2016].   Pearse's site is based on M.J. Vermaseren's *Corpus Inscriptionum et Monumentorum Religionis Mithriacae* (2 vols, The Hague: Martinus Nijhoff, 1956-60), with the bonus that monuments discovered since the publication of Vermaseren's *Corpus* are appended at the end of the website in a supplement. 'Monuments' (i.e., mithraea and their contents) in Vermaseren's *Corpus* will be referenced by number with the prefix 'V'. Numeration there is by Roman province/region and is continuous from the first to the second volume. Italian monuments are all in Vol. 1.
[5] Porphyry, *On the Cave* 24.
[6] On Cautes and Cautopates see John R. Hinnells, 'The Iconography of Cautes and Cautopates', *Journal of Mithraic Studies* 1, no. 1 (1976): pp. 36–67.

Cautopates, a lowered torch. They represent principles of cosmic polarity and opposition. Thus, as is universally agreed, Cautes represents the rising Sun, Cautopates the setting Sun, both in relation to Mithras as the Sun culminating at noon.[7] Of interest here is the fact that, while there are scores of extant visual representations of them in sculpture and occasionally in fresco and numerous mentions of them in epigraphy, only in Porphyry *On the Cave* 24 do they appear in a literary source contemporary with the Mysteries.

At Sette Sfere the pair are represented in mosaic on the ends of the benches nearest to the entrance. Cautes with his raised torch is on the right bench end, Cautopates with his lowered torch on the left.[8] Thus, from the point of view of the tauroctonous Mithras in the cult-niche, Cautes is on the left on the bench with the southern signs and Cautopates on the right on the bench with the northern signs. This is precisely the relationship that Porphyry describes in the passage above. Microcosm matches macrocosm. Text matches monument.

Very often the torchbearers are present in the tauroctony, flanking the scene, one on the (viewer's) right at the bull's head, the other on the left at the bull's tail. However, their positions are interchangeable. Sometimes Cautes is on the right and Cautopates on the left, sometimes vice versa. This interchangeability is the single regular variation seen in the composition of the tauroctony, and it appears to have followed regional norms.[9] In contrast, in the small minority of instances where the torchbearers are found on or next to the benches, one each side, they are

---

[7] Cautes also represents the waxing Sun, increasing daily in altitude at the zenith and in astronomical latitude, and Cautopates the waning Sun, decreasing daily in both of those regards. See Roger Beck, 'Cautes and Cautopates: some astronomical considerations', *Journal of Mithraic Studies* 2, no. 1 (1977): pp. 1–17. Reprinted in Roger Beck, *Beck on Mithraism* (Aldershot: Ashgate Publishing, 2004), pp. 133-39.

[8] V243 (illustrations in http://www.ostia-antica.org/regio2/8/8-6.htm [accessed 21 March 2016]). For a pair of statues of Cautes and Cautopates, also from an Ostian mithraeum Mitreo del Palazzo Imperiale, see V254 (cf. http://www.ostia-antica.org/regio1/pi/pia.htm [accessed 21 March 2016]).

[9] Hinnells, 'Cautes and Cautopates', pp. 36–67. For an example of a tauroctony with Cautes on the left see the fresco of the Barberini Mithraeum (V390): http://roger-pearse.com/mithras/display.php?page=cimrm390 [accessed 6 May 2016]. For an example of a tauroctony with Cautes on the right see the obverse of the Heddernheim relief (V1083): http://roger-pearse.com/mithras/display.php?page=cimrm1083 [accessed 6 May 2016].

invariably positioned as at Sette Sfere: Cautes on the right (as one enters), Cautopates on the left.[10]

This is the point at which to confront the objection already intimated at the end of Part 1 of this article.[11] Sette Sfere is but one mithraeum among scores the length and breadth of the Roman Empire: are there many like it, and if not, why treat it as exemplary? It is indeed true that there is no other extant mithraeum quite like Sette Sfere. So a first and tentative answer might be that Sette Sfere merely makes explicit what was implicit in other mithraea. This answer was given as long ago as 1976 by Richard Gordon.[12] The fit between Sette Sfere and Porphyry's description of what he presents as a generic 'cave' goes some way towards validating Gordon's view. In other early articles, Gordon speaks of Mithraic imagery as fluctuating between 'poles of "silence" and "garrulity"'.[13] On this continuum, Sette Sfere would be a very 'garrulous' monument, explicit about what is implicit in other mithraea.

This argument retains some validity, even though those in the field no longer think in terms of a set Mithraic doctrine current in the empire *semper et ubique*.[14] What I am going to suggest is that Sette Sfere represents, in the most explicit form, a stream of Mithraism current at least in west central Italy (Campania, Latium, including the city of Rome and its port, Ostia, and southern Etruria) in the second and third centuries CE. It was this stream of Mithraism, I suggest, that was known to Porphyry, whether directly or through intermediaries.[15]

---

[10] Leroy A. Campbell, *Mithraic Iconography and Ideology* (Leiden: Brill, 1968), p. 42.

[11] Beck, MMU Part 1.

[12] Richard Gordon, 'The sacred geography of a mithraeum: the example of Sette Sfere', *Journal of Mithraic Studies* 1, no. 2 (1976): pp. 119–65. Reprinted in Richard Gordon, *Image and Value in the Graeco-Roman World: Studies in Mithraism and Religious Art*, Variorum Collected Studies Series, CS551 (Aldershot: Ashgate Publishing, 1996), Ch. VI (no repagination) [hereafter Gordon, *Image and Value*].

[13] See Gordon, *Image and Value*, index under 'garrulity, of images'.

[14] See Roger Beck, *The Religion of the Mithras Cult in the Roman Empire: Mysteries of the Unconquered Sun* (Oxford: Oxford University Press, 2006), pp. 41–64.

[15] Porphyry mentions by name Cronius, Numenius, Pallas, and Eubulus. On these lost sources see Robert Turcan, *Mithras Platonicus: Recherches sur l'hellénisation philosophique de Mithra* (Leiden: Brill, 1975).

What distinguishes the mithraea of west central Italy is the pair of mid-bench niches present in a number of them, but (to my knowledge) absent from mithraea in the rest of the Empire. This pair of niches, I argue, represents the solstices, the summer solstice in or at the start of Cancer and the winter solstice in or at the start of Capricorn. As well as Sette Sfere (V239),[16] they are found in the Ostian Mitreo del Palazzo Imperiale (V250),[17] and in Rome in the Barberini Mithraeum (V389) and the Mithraeum in the Baths of Caracalla (V457); to the north of Rome, in the Spoleto Mithraeum (V673) and the Vulci Mithraeum;[18] and to the south in Campania, in the Capua Mithraeum (V180). Granted, this is a small fraction of the known mithraea in the region (six of fewer than thirty), but it is not insignificant when contrasted with the complete absence of paired mid-bench niches from mithraea so far discovered elsewhere in the Empire. I am not saying that the equation of *the mid points of the benches* with the solstices was unknown outside of west central Italy – it follows, after all, from the mithraeum's cosmography –only that in this area it was occasionally emphasized and made explicit by *the opposed pair of niches*.

Next we must confront a paradox in the positioning of the torchbearers. It will be recalled that in the comparatively few instances when the torchbearers are *not* part of the tauroctony, Cautes with his raised torch is always on the right as one enters (to the left of the bull-killing Mithras in the cult-niche) and Cautopates on the left (to Mithras's right).[19] Why, though, do we find the torchbearer with the *lowered* torch, Cautopates, on the side of the mithraeum associated with *summer* (summer solstice, signs

---

[16] See above, footnote 4, for the significance of the catalogue numbers prefixed with 'V'.

[17] It was in the Mitreo del Palazzo Imperiale that the statues of Cautes and Cautopates (V254/5), already mentioned, were set on bases in or just in front of the niches, Cautes on the right, Cautopates on the left. See http://www.ostia-antica.org/regio1/pi/pia.htm [accessed 21 March 2016].

[18] For plans and pictures of the Vulci Mithraeum, to which we shall return at the conclusion of this article, see Anna Maria Sgubini Moretti, 'Nota preliminare su un mitreo scoperto a Vulci', in U. Bianchi, ed., *Mysteria Mithrae* (Leiden: Brill, 1979), pp. 260–95. There are illustrations of this mithraeum and its finds, together with a discussion, in:
http://roger-pearse.com/mithras/display.php?page=supp_Italy_Vulci_Mithraeum [accessed 21 March 2016].

[19] The figures of Cautes and Cautopates are not always at or on the *ends* of the benches, as at Sette Sfere. At Palazzo Imperiale they were in or just in front of the mid-bench niches. See http://www.ostia-antica.org/regio1/pi/pia.htm [accessed 21 March 2016].

of the zodiac *above* the celestial equator), and the torchbearer with the *raised* torch on the side associated with *winter* (winter solstice, signs of the zodiac *below* the celestial equator)? It seems – and is – counterintuitive, since the torchbearers represent opposite aspects of the *solar* deity Mithras. Why, then, place Cautopates, the avatar of the *low* sun, on the *summer* side of the mithraeum, among the signs through which the sun journeys (both spatially and temporally) from the spring to the autumn equinox, and Cautes, the avatar of the *high* sun, on the *winter* side among the signs through which the sun journeys back again from the autumn to the spring equinox?

In the classical world, some people believed not only that human souls descended to mortality on Earth through the seven planetary spheres and returned back through them to heaven after death, but also that they entered and started their descent through a gate in the sphere of the fixed stars located at the summer solstice at the start of Cancer and left after their ascent back up again through a diametrically opposite gate at the winter solstice at the start of Capricorn.[20] Among those holding this belief were the Mithraists. We know from Porphyry that 'they initiated their members into a mystery of the soul's descent and return back out again'.[21] We know that they designed and constructed their mithraea as 'models of the cosmos's[22] so as to enable this mystery. Finally, we are finding in actual mithraea strategically located symbols of the solstices.

The conclusion is inescapable. For the Mithraists the mid-bench niches, which are the solstices, are also, cognitively and for ritual purposes, the gates through which their souls descend into mortality and ascend back out again into immortality, the gates of genesis and apogenesis. To cross the mithraeum from one side to the other at the mid-point of the aisle, whether in imagination or as a ritual step, is fraught with meaning.[23] It is nothing less than cosmic soul travel.

And now we know why Cautopates is always placed on the 'northern' bench and Cautes on the 'southern'. It is because Cautopates presides over *descent* down into mortality from the summer solstice at the northernmost

---

[20] See Roger Beck, *Planetary Gods and Planetary Orders in the Mysteries of Mithras* (Leiden: Brill, 1988), pp. 93-96. See also Beck, *Religion of the Mithras Cult*, pp. 129-30. Of the ancient sources, the most important and extensive is Porphyry, *On the cave* 21-29.

[21] Porphyry, *On the cave* 6.

[22] Porphyry, *On the cave* 6.

[23] For an echo of such a ritual in the Mithras cult see Beck, *Religion of the Mithras Cult*, pp. 129-30.

point on the ecliptic, while Cautopates presides over *ascent* back up to the winter solstice at the southernmost point on the ecliptic and so into immortality. Ultimately it is Mithras, 'the creator and lord of genesis', from his 'proper throne at the equinoxes', who rules both processes and both journeys.[24]

And so to the planets, those powerful yet erratic beings who occupy the spheres between Earth and the ultimate heaven of the fixed stars, spheres through which the human soul must travel both descending to mortality and rising back up to immortality.[25]

In the Mithraeum of the Seven Spheres the planets are represented in two forms: first as an undifferentiated sequence of seven arcs, executed in black mosaic, extending up the central aisle; secondly as six – not seven – differentiated, anthropomorphic beings, again on black-on-white mosaic, set three aside on the faces of the side-benches. The precise placement of the two sets may be seen in Figure 2. Easy to interpret are the seven arcs on the floor. Here, surely, are the planetary spheres through which the soul descends and returns. To pace up or down the central aisle is thus to replicate the journey of souls.

There is another Ostian mithraeum that has, in mosaic, both anthropomorphic images of six of the seven planets and a design of seven arches. This is the 'Mithraeum of the Seven Gates' or 'Sette Porte' (V287).[26] It is the seven arches that immediately concern us (V288.1).

---

[24] Porphyry, *On the cave* 24.

[25] Were my study primarily about the Mithraic soul journey rather than about the mithraeum as a model universe, this would be the point at which to introduce the testimony of Origen, *Contra Celsum* (*Against Celsus*) 6.22: 'These things [i.e. the celestial ascent of souls] the teaching of the Persians [i.e. the Mithraists] and their initiation into the Mysteries of Mithras intimate.... for they have a symbol of the two celestial revolutions (*periodôn*), that of the fixed stars and that assigned to the planets, and of the route of the soul through and out (*diexodou*) of them. The symbol is this: a seven-gated ladder and an eighth [sc. gate] on top'. (We would probably speak of a ladder with seven-plus-one 'steps' or 'rungs'.) My point is precisely that Origen is talking about the symbol of a ladder and *not* about the symbol of the mithraeum. On this passage from Origen, see Beck, *Planetary Gods*, pp. 73-77; also Beck, *Religion of the Mithras Cult*, pp. 83 f. Interestingly, the order of the planets/gates is temporal, not spatial; it is the order of the days of the week in reverse, from lowest (Saturday/Saturn) to highest (Sunday/Sun). For Origen's text see Henry Chadwick's translation with introduction and notes: Origen, *Contra Celsum* (Cambridge: Cambridge University Press, 1953), p. 334.

[26] For a description and illustrations of Sette Porte, see: http://www.ostia-antica.org/regio4/5/5-13.htm [accessed 21 March 2016].

Truth to tell, '*porte*'/'gates' is a somewhat tendentious description. What is represented is rather an arcade with a central gateway and arches, or more precisely openings, between the pillars, three on either side. But let us accept, as everyone does, that they represent gates through the spheres of those same planets which are represented anthropomorphically in the same mithraeum.

There are several important points to make about this row of seven 'gates'. First, it runs *across* the aisle, just inside the entrance, whereas the sequence of arcs at Sette Sfere runs *up* the aisle from entrance to cult niche. However, this does not mean that one or other of the designers 'got it wrong', but rather that, arguably, there is more than one way of 'getting it right', in other words more than one way of setting the 'symbols of the elements and climates of the universe' 'in proportionate arrangement'.[27] The Sette Porte arrangement across the aisle has the advantage of book-ending this row of seven planetary 'gates' with the images of the torchbearers, who represent descent and ascent, facing each other on the fronts of the benches at the entrance end of the mithraeum. From the remains of their images it is impossible to tell which figure carries the raised torch and which the lowered. However, as we noted above, in all other mithraea where the torchbearers are *not* part of the tauroctony, Cautes with his raised torch is always on the right as one enters and Cautopates on the left. We have no reason to doubt that the same placement holds here at Sette Porte.

The central 'gate', as described above, is much larger than the three 'gates' on either side of it. If we 'read' the planetary gates in the conventional astronomical order of distance outward from the earth or inward from the sphere of the fixed stars, then, most appropriately, the large central archway would represent the gate of the Sun, and the other arches, from left to right, the gates of the other planets as shown in Figure 3.

Note that to move across the aisle here at the entrance, whether physically or 'in the mind's eye', is to *descend* through the planetary gates when moving left to right or to *ascend* when moving right to left.

---

[27] Porphyry, *On the cave* 6.

(A)

Left bench   right bench

↑

Up aisle to cult niche

CAUTOPATES   Cautes

    Saturn   Jupiter Mars   Sun   Venus Mercury Moon

Descent to earth ───▶

    Gate 1   Gate 2   Gate 3 Gate 4 Gate 5 Gate 6 Gate 7

             Entrance to mithraeum

(B)

Left bench   right bench

↑

Up aisle to cult niche

Cautopates   **CAUTES**

    Saturn Jupiter   Mars   Sun Venus Mercury Moon

◀─── Ascent to heaven

    Gate 7   Gate 6   Gate 5 Gate 4 Gate 3 Gate 2 Gate 1

             Entrance to mithraeum

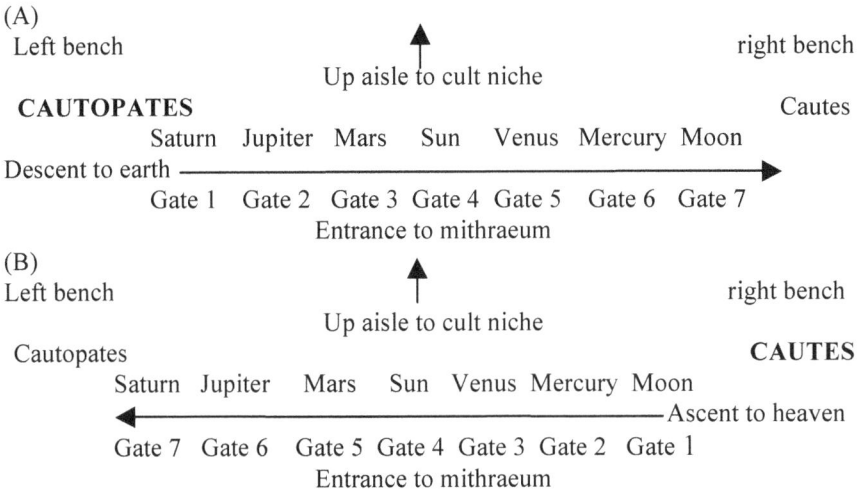

Fig. 3. Interpretation of the mosaic of the seven 'gates' on the floor at the entrance to the mithraeum of that name ('Sette Porte') in Ostia. (A) Read from the left = descent from heaven to earth (Cautopates presiding). (B) Read from the right = ascent from earth to heaven (Cautes presiding).

Let us return to Sette Sfere and the disposition of the images of six of the seven planets on the front of the benches, three on each side.[28] The first question we must ask is why there is no image of the Sun, the seventh planet and arguably the most important, especially in a solar cult. The answer of course is that the Sun's image is not absent at all. It is present as the image of the tauroctonous Mithras, himself the Sun, at the top of the aisle. That location in the mithraeum's celestial layout is the spring equinox at the start of Aries. As we saw above, this is precisely the 'proper seat' of Mithras 'at the equinoxes' with the northern signs to his right and the southern to his left.[29] This is an *ideal* position, occurring as an *actual* position once a year on the day of the spring equinox.

Is the proximity of the images of the other six planetary images to the images of certain signs meaningful? Proceeding counter clockwise around the benches in the order of the signs of the signs of the zodiac (see Fig. 2), we find that:

---

[28] See http://www.ostia-antica.org/regio2/8/8-6.htm (final images) [accessed 6 May 2016].

[29] Porphyry, *On the cave* 24.

the image of Jupiter is closest to the image of Aries,
the image of Mercury is closest to the image of Gemini,
the image of the Moon is closest to the image of Virgo,
the image of Mars is closest to the image of Libra,
the image of Venus is closest to the image of Capricorn,
and the image of Saturn is closest to the image of Pisces.

Why this sequence and why these correlations? To answer these questions, we must begin with another, prior question: what are the implications of placing images of the planets on the side-benches at all? Now the side-benches together represent the circle of the zodiac. It follows that placing planetary images on the benches makes a statement about their positions on the zodiac relative to each other at a point in time.

In this context the proportionate arrangement of cosmic symbols means something entirely different from what it has meant for us so far. We are no longer talking about fixed sequences and celestial dispositions which obtain permanently. Rather, we are talking about highly fluid sequences and dispositions which may – or may not – obtain today but certainly did not obtain last week and will not obtain next week. To appreciate the difference, consider, on the one hand, the immutable order of the planets in depth of space and, on the other hand, their constantly changing relationships, each to the other two, in longitude around the ecliptic. It is the latter to which an arrangement of planetary symbols on the side benches necessarily refers.

Let us quickly review the options. (1) The arrangement could be actual, in the sense of recording the actual disposition of the planets relative to the Sun at the time of the spring equinox in a particular year when the mithraeum was in service. Or (2) it could be an ideal arrangement, a configuration which *ought* to obtain in the heavens or perhaps what *will* obtain at the End of Time. Or (3), as a blend of the two alternatives, it may be that at the time of the spring equinox in a particular year an especially favourable configuration *did* obtain.

First, it can be said with certainty that what is represented at Sette Sfere is not a horoscope in the straightforward sense that the planets on the given date were actually in the signs of the zodiac closest to which they are shown on the side-benches. The inferior planets Mercury and Venus, as a matter of brute astronomical fact, cannot be the distance in longitude away from the Sun which the placement of their images would suggest. Mercury is never further than one sign away from the Sun, and Venus never further than two. It follows that when the Sun is at the spring equinox in or at the start of Aries, Mercury cannot be at least two signs away in Gemini or Venus in Capricorn three signs away (see Fig. 2).

If Sette Sfere does commemorate an historical planetary configuration, it can only be one in which the planets were disposed to the east and west of the equinoctial Sun in the order shown, but not at the longitudes implied. What we would be looking for would be a spring equinox at the time of the time of Sette Sfere's construction – more precisely, at the time of the installation of the floor and bench mosaics –when the superior planets Jupiter and Saturn were closer in longitude to the Sun than *both* of the inferior planets Mercury and Venus, since that is the configuration signalled at Sette Sfere: Jupiter in Aries between the Sun, also in Aries, and Mercury in Gemini; and Saturn in Pisces between the Sun in Aries and Venus in Capricorn.

Sette Sfere dates from the 'second half of the second century A.D.' (V239), and nothing suggests that the mosaics were a later addition. The postulated configuration (the two superior planets both closer to the Sun than the two inferior planets) did in fact obtain at the spring equinox of 172 CE:[30]

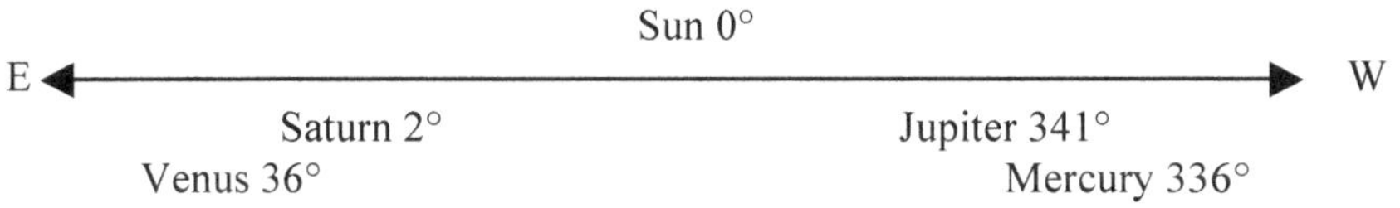

Sun 0°

E ←————————————————————————————————————→ W

Saturn 2°   Jupiter 341°
Venus 36°   Mercury 336°

Fig. 4. Longitudes of Mercury, Jupiter, Sun, Saturn, and Venus at the spring equinox on 21 March 172 CE.

If this configuration was indeed what the Sette Sfere designers were commemorating, one must admit that they got it partly wrong. They should have put Saturn and Venus on the left-hand 'northern' bench and Jupiter and Mercury on the right-hand 'southern' bench, not vice versa (compare Figures 2 and 4). But that of course would have meant reversing the direction of the order of the zodiac around the mithraeum, which would in turn have compromised the correspondences of microcosm to macrocosm through which the mithraeum functions as an accurate 'image of the universe' by 'proportionate arrangement' of (literally) 'the things inside'.[31]

---

[30] For what follows, see Roger Beck, 'Sette Sfere, Sette Porte, and the spring equinoxes of A.D. 172 and 173', in Ugo Bianchi, ed., *Mysteria Mithrae* (Leiden: Brill, 1979), pp. 515-29.

[31] Porphyry, *On the cave* 6.

Perhaps we should entertain a different solution which retains the reference to the actual celestial configurations of those times while respecting the integrity of the mithraeum's symbolic design.

Astrologically, the actual planetary configuration on 21 March 172, replicated (albeit with some error) in the configuration at Sette Sfere, instantiates the celestial relationship of *doryphoria*.[32] 'Doryphoria' means literally 'spear-carrying' in Greek. To 'carry a spear' for someone means to escort them in a procession. And that is precisely what the planets Venus and Saturn (in front) and Jupiter and Mercury (behind) did for the actual phenomenal Sun in the heavens on 21 March 172, and for Mithras the Unconquered Sun God (at his proper seat at the spring equinox) in the microcosm of the Mithraeum of the Seven Spheres in Ostia (again, compare Figures 2 and 4).

Let us now return to the Mithraeum of the Seven Gates, where there is also an arrangement of images of six of the seven planetary gods in mosaic. However, at Sette Porte there is no arrangement of images of the signs of the zodiac, as there is at Sette Sfere. Consequently, there can be no question of particular planets being 'in' particular signs – which even at Sette Sfere we saw was a somewhat misleading trail. At Sette Porte, dated 160-170 CE (V287), the images of the planets are arranged thus:[33]

	(Sun 0°)	
	Saturn 4°	
	Jupiter 9°	
Mercury 349°		Moon 239°
Venus 3°		Mars 279°
on the left bench	on the aisle	on the right bench

Fig. 5. Disposition of the images of the planetary gods in the Mithraeum of the Seven Gates (Sette Porte), Ostia, with longitudes on the spring equinox, 21 March 173.

I have added the longitudes of the planets on the date of the spring equinox of 173 CE, rather than 172. This I have done to demonstrate that Sette Porte in all likelihood commemorates *doryphoria* of the Sun by the two

---

[32] On *doryphoria* in astrology, see Auguste Bouché-Leclercq, *L'astrologie grecque* (Paris: E. Leroux, 1899; repr. Brussels: Culture et Civilization, 1963), pp. 252-4.
[33] For illustrations of these images, see http://www.ostia-antica.org/regio4/5/5-13.htm [accessed 6 May 2016].

superior planets, represented in the aisle closest to Mithras as Sun in the cult-niche, in the year following commemoration at Sette Porte.

The question remains: how would the Mithraists of Sette Sfere and Sette Porte have known about these configurations of *doryphoria* in the two years in question? One answer is that they could have ascertained imminent *doryphoria* from tables of planetary longitude. Indeed they could – but there is a simpler answer which does not postulate access to astrologers' tables. Close *doryphoria* to the Sun is a matter of what you do *not* see, or rather, *no longer* see. As the date of the spring equinox in each of these years approached, one would have seen Jupiter and Saturn disappear from the post-sunset western sky, not to reappear in the pre-dawn eastern sky until some weeks later.[34]

172 CE	173 CE
Saturn	*Saturn*
last evening visibility: end of Feb	last evening visibility: mid March
first morning visibility: mid May	first morning visibility: mid May
Jupiter	*Jupiter*
last evening visibility: mid Feb	last evening visibility: early March
first morning visibility: late April	first morning visibility: mid May

Fig. 6. Approximate dates of last evening visibility and first morning visibility of Saturn and Jupiter at Ostia in 172 and 173 CE

It was during the periods of invisibility so defined that the two senior planets would be known to be 'accompanying' the equinoctial Sun. The observations are of the easiest, and the requisite knowledge of apparent planetary motions of the most basic. As we would say today, 'it ain't rocket science'.

To conclude, we return to the Vulci Mithraeum in Etruria.[35] This mithraeum is distinguished by exceptionally high side-benches. On each side the bench is carried on an arcade of six arches, with a niche in the middle and niches at both ends. The obvious inference to draw is that the twelve arches represent the signs of the zodiac, essentially as at Sette Sfere, only with arches substituted for mosaic images of the signs. Again, as at

---

[34] In order to estimate the dates, I reproduced the celestial phenomena observable from Ostia over the given time period by means of the 'Voyager 4.5 Dynamic Sky Simulator' software program.

[35] For references, see above, n. 16.

Sette Sfere the mid-bench niches represent the solstices.[36] The niches at the bench ends, I suggest, represent the equinoxes, the two at the cult-niche end representing the spring equinox at the end of Pisces and the beginning of Aries, and the two at the entrance end representing the autumn equinox at the end of Sagittarius and the start of Capricorn. These bench-end niches are *explicit signifiers* of the equinoxes, which are *precisely located,* in the microcosm of the mithraeum, at the cult-niche (spring equinox) and the entrance (autumn equinox).

cult-niche
= spring equinox
[bull-killing Mithras]

bench to the left	bench to the right
i.e. 'to Mithras's right':	i.e. 'to Mithras's left'
niche = spring equinox	niche = autumn equinox
northern signs of the zodiac	**southern signs**
(1) **spring quadrant**	(4) **winter quadrant**
arch = Aries	arch = Pisces
arch = Taurus	arch = Aquarius
arch = Gemini	arch = Capricorn
mid-bench niche	**mid-bench niche**
= summer solstice (Cancer)	= winter solstice (Capricorn)
'genesis'	'apogenesis'
= 'descent of souls'	= 'exit [sc. of souls] back out again'
(2) **summer quadrant**	(3) **autumn quadrant**
arch = Cancer	arch = Sagittarius
arch = Leo	arch = Scorpio
arch = Virgo	arch = Libra
[Cautopates]	**[Cautes]**

entrance
= autumn equinox

Fig. 7. Design of the Vulci Mithraeum.

By placing movable images of the planets in the appropriate zodiacal niches, any configuration of the heavens, past, present, or future, actual or ideal, could be replicated in the Vulci Mithraeum. Of all mithraea, Vulci is thus the most adaptable 'model of the universe'[37] so far discovered.

---

[36] At Vulci, however, these niches are differentiated by shape, that on the left, representing the summer solstice, being rectangular in cross-section, that on the right, representing the winter solstice, arciform.
[37] Porphyry, *On the cave* 6.

The Interpretation of a Horoscope Cast by Abraham the Jew in Béziers for a Child Born on 29 November 1135: An Essay in Understanding a Medieval Astrologer

Helena Avelar, with an introduction by Charles Burnett

Abstract: This article analyses the horoscope written by a certain 'Abraham Judeus' at Béziers for a child born in 1135. It shows that, given the positions of the heavenly bodies and the astrologer's interpretation, one can work out how the astrologer arrived at his judgments, using the standard astrological doctrine of the time.

Introduction

A twelfth-century manuscript in Paris, Bibliothèque nationale de France, lat. 16208, contains on its flyleaves a collection of notes, including several horoscopic charts.[1] The body of the manuscript contains texts on astrology, astronomy, the astrolabe, and arithmetic, several of which were authored by Raymond of Marseilles, an astronomer based in Marseilles in the mid-twelfth century (fl. 1141).[2] The manuscript eventually came into the hands of (if it was not copied for) Roger of Fournival, doctor and astrologer to the French kings Philippe-Auguste and Louis VIII, who added most of the notes, including the horoscope of Philippe-Auguste and a horoscope of his

---

[1] For a full description of the manuscript see David Juste, *Catalogus codicum astrologorum latinorum II, Les manuscrits astrologiques latins conservés à la Bibliothèque nationale de France à Paris,* (Paris: CNRS Éditions, 2015), pp. 236–240.

[2] These include his *Liber iudiciorum*, and fragments of his *Liber cursuum planetarum* and treatise on the astrolabe: see Raymond de Marseille, *Opera omnia I: Traité de l'astrolabe, Liber cursuum planetarum*, ed. Marie-Thérèse d'Alverny, Charles Burnett, and Emmanuel Poulle (Paris: CNRS editions, 2009), pp. 40–42.

Helena Avelar, with an introduction by Charles Burnett, 'The Interpretation of a Horoscope Cast by Abraham the Jew in Béziers for a child born on 29 November 1135: An Essay in Understanding a Medieval Astrology', *Culture and Cosmos,* Vol. 18, no. 2, Spring/Summer 2014, pp. 19-40.
www.CultureAndCosmos.org

own birth (6 May 1179).[3] It then became part of the famous library of Roger's son, Richard of Fournival, and was bequeathed by him to Gérard d'Abbéville, and thence to the library of the Sorbonne.[4] The only horoscope with an interpretation occurs on the recto of the initial flyleaf. This is a horoscope cast by Abraham the Jew in Béziers for a child born on 29th November 1135.

This horoscope has been edited and examined before.[5] But up to now no one has tried to work out how the astrologer came to the conclusions that he reaches in his text. It is especially interesting to do this because the 'Abraham Judeus' who cast the horoscope is likely to be Abraham Ibn Ezra, whom we know to have been in Béziers from other evidence.[6] Ibn Ezra wrote two (or even three) sets of texts on astrology, one in Béziers in 1148, and some parts of the interpretation can be compared with the instructions given in the versions of the *Book of Nativities* (*Sefer ha-Moladot*) that formed part of these sets.[7] Helena Avelar shows how the prognostics in this horoscope follow naturally from the astrological rules of Ibn Ezra, which were typical of his time. For as Abraham Ibn Ezra himself draws upon the astrological doctrines of Ptolemy, Dorotheus, Māshā'allāh, Abū Ma'shar, Sahl ibn Bishr and al-Qabīsī, his doctrines can be taken as representative of these doctrines. After the title of the horoscope (within

---

[3] See Emmanuel Poulle, 'La date de naissance de Louis VIII', *Bibliothèque de l'École des chartes* 143 (1987): pp. 427–430.
[4] Richard of Fournival, *Biblionomia*, in Leopold Delisle, *Le cabinet des manuscrits de la Bibliothèque impériale* (Paris: Bibliothèque impériale, 1826–1910, Vol. 2.
[5] Joshua Lipton, 'The Rational Evaluation of Astrology in the Period of Arabo-Latin Translations, ca. 1126-1187 AD', (unpublished PhD dissertation, University of California, Los Angeles, 1978), pp. 221–222. Lipton corrected the date '1136' to '1135'. See also John D. North, *Horoscopes and History* (London: The Warburg Institute, 1986), pp. 108–9, who compares the planetary positions in the manuscript with those worked out by modern techniques, and suggests that the latitude indicates that the place of birth could have been Tarragona.
[6] Shlomo Sela and Gad Freudenthal, 'Abraham Ibn Ezra's Scholarly Writings: A Chronological Listing', *Aleph* 6 (2006): pp. 13–55.
[7] *Abraham Ibn Ezra on Nativities and Continuous Horoscopy: a Parallel Hebrew-English Critical Edition of the Book of Nativities and the Book of Revolution*, edited, translated, and annotated by Shlomo Sela (Leiden: Koninklijke Brill NV, 2014), henceforth referred to as *Nativities*.

the horoscopic figure) the interpretation is analysed sentence by sentence, with Abraham the Jew's words printed in bold.[8]

The Text

1. *Nativitas cuiusdam pueri anno domini .m.c.xxxvi., .xxix. die Octobris, die Martis, hora .ii.a.*

The nativity of a boy born in the year 1135, on the second hour of the 19th day of October, a Tuesday).

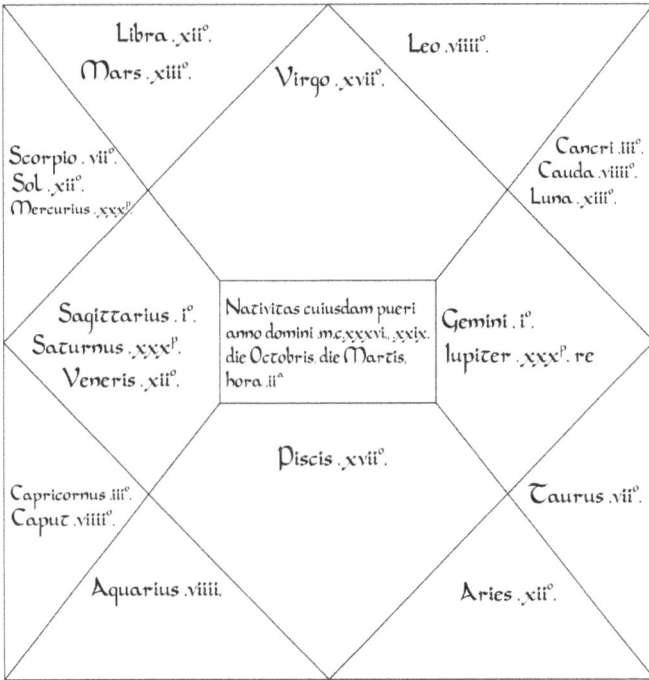

Fig. 1. Reconstruction of the horoscope (by Luís Ribeiro).

Chart calculated for 29th October 1135, at 8:33 (by the very end of the hour of Venus, the second planetary hour), in Tarragona, Spain (41°N07' 001°E15'). The reconstruction used the computer program *Solar Fire Pro*.[9]

---

[8] This is a corrected version of the transcription in Lipton, 'The Rational Evaluation of Astrology in the Period of Arabo-Latin Translations', pp. 221–222. Lipton corrected the date '1136' to '1135'.

[9] Note that the hypotheses presented are based on the writings of Abraham ibn Ezra in *Sefer ha-Moladot* (*Book on Nativities*) and *Sefer ah-Tequfah* (*Book of*

Birth time calculated in LAT (Local Apparent Time: -0:20:27).

The comparison between the horoscope calculated by Abraham and the one calculated by computer reveals minor variants in the houses' cusps and in the position of the planets. These are due mainly to Abraham's notation, which uses only the degrees and omits the minutes, and not to significant errors of calculation. The greatest difference is in the position of Mercury: 30° of Scorpio when it should be 23° of Scorpio.[10] The position of Saturn, correctly placed at 30° Sagittarius but above Venus, which is at 12° Sagittarius, is an error of drawing, not of calculation.

	Horoscope calculated by Abraham	Horoscope calculated by computer
House I	Sagittarius 01° Saturn – Sagittarius 30° Venus – Sagittarius 12°	Sagittarius 01°45' Venus – Sagittarius 11°45' Saturn – Capricorn 00°58'
House II	Capricorn 03° North Node – Capricorn 09°	Capricorn 05°31' North Node – Capricorn 08°50'
House III	Aquarius 09°	Aquarius 09°56'
House IV	Pisces 17°	Pisces 17°49'
House V	Aries 12°	Aries 13°30'
House VI	Taurus 7°	Taurus 08°25'
House VII	Gemini 01° Jupiter – Gemini 30° Retrograde	Gemini 01°45' Jupiter – Gemini 29°11' Retrograde
House VIII	Cancer 03° South Node – Cancer 09° Moon 13°	Cancer 05°31' South Node – Cancer 08°50' Moon 13°12'
House IX	Leo 09°	Leo 09°56'
House X	Virgo 17°	Virgo 17°49'
House XI	Libra 12° Mars – Libra 13°	Libra 13°30' Mars – Libra 14°03'
House XII	Scorpio 07° Sun – Scorpio 12° Mercury – Scorpio 30°	Scorpio 07° Sun – Scorpio 12°01' Mercury – Scorpio 23°41'

Fig. 2. Table I – Comparing calculations.

The first house, signifying the native himself, is the one that receives most

---

Reasons). These are sometimes compared with the doctrines in *Al-Qabīṣī (Alcabitius): The introduction to Astrology: editions of the Arabic and Latin texts and an English translation*, ed. C. Burnett, K. Yamamoto, and M. Yano (London: The Warburg Institute, 2004), p. 55.
[10] The 30th degree of a sign corresponds to the 1st degree of the next sign. In this case, 30° of Scorpio can also be represented as 0° Sagittarius.

attention in the interpretation. It describes the native's temperament, complexion, intelligence, manners and ethics – in short, it comprises what is presently called the native's 'psychological traits'.[11] It also includes the calculation of the native's life span.

The other eleven houses describe the native's circumstances: resources, family, children, friends, and all the other matters of his life.

House	Meanings (abridged version)
First House	The native (also the soul, knowledge, intelligence and belief, the body and life)[12]
Second House	Wealth, helpers
Third House	Brothers, relatives, familiar wisdom[13]
Fourth House	The father, buried treasures, the end of everything; landed property
Fifth House	Sons, food, drink and clothing
Sixth House	Hidden wars, like illnesses and deformities
Seventh House	The female (the wife), the partners, wars
Eighth House	Death
Ninth House	Journeys, wisdom, belief
Tenth House	The mother; dominion, greatness; the art on account of which the native is known
Eleventh House	Lovers, beauty and honour
Twelfth House	Quarrels, dishonour, shame, prison; animals that men ride on (large-size animals)

Fig. 3. Table II – The astrological houses according to Ezra.[14]

Some topics pertaining to the other houses are somehow connected to the ones from the first and appear together in the interpretation. Such is the case of religion, a topic from the ninth house, which is sometimes integrated into the interpretation of the native's mentality, and therefore

---

[11] The concept of complexion comprises both physical appearance (aspect, vitality, proneness to certain illnesses and so on) and psychological traits (temperament, character, intelligence and proneness to mental problems).

[12] S. Selam ed., *Abraham ibn Ezra, The Book of Reasons* (Leiden: Brill, 2007), p. 99 and pp. 211–213 [hereafter *Book of Reasons*].

[13] Familiar wisdom: 'the body of knowledge that is close to one's law or religion'; in this case, the wisdom of the Torah; *Book of Reasons*, p. 153.

[14] *Book of Reasons*, pp. 67–69 and pp. 215–217 (for the colours of the houses). See also *Nativities*, pp. 99–183.

presented as part of the first-house interpretation.[15]

It is worth noticing the complex use of the planets in interpretation, combining their general significance and their specific meaning in a given horoscope (whether by house position or by house rulership), according to the context of the topic being scrutinized.

For instance, Venus is a general significator of women and love in any horoscope, and it is frequently taken as part of the description of the native's wives; but in this horoscope it is placed in the first house, so it also has some part on the native's personality and manners; moreover, it is also the ruler of the sixth and eleventh houses, and can therefore contribute to the interpretation of this native's health and servants (sixth house matters) and friends (eleventh house matters).[16] Similarly, Jupiter is a general significator of benefice and honour, so it is always part of the interpretation for the native's wealth and social recognition; but in this case Jupiter is placed in the seventh house, which makes it an active part of the interpretation for marriage; it is also the ruler of the Ascendant and the fourth house, which makes it a significator for the native himself (Ascendant) and his family and origins (fourth house).[17]

2. *Aspexi hanc nativitatem et inveni quod pre omnibus parentibus suis erit sapientior et erit amator scientie et correctionis et diliget sapientes; et erit mente benevolus et libenter suae rogabit.*

I have inspected this nativity and found that the child will be wiser than all his relatives and will be a lover of science and of correcting and he will love (the company of) wisemen; he will be intellectually (*mente*) generous and will freely give of his own.

This is a typical part of the interpretation for the first house, signifying the native, his temperament and complexion, his mind and humours, and all his natural inclinations.

In this chart there is an extended trine between Mercury and the Moon, an indication of a positive mind:

---

[15] Ibn Ezra starts his description of the first house stating that it signifies 'the soul, knowledge, intelligence and belief': *Nativities*, p. 99. Also relevant is Sela's comment on page p. 233–234: 'It is odd to find "belief" in this list of the indications of the first place, because it fits the indications of the third place or the ninth place, but certainly not the first place. (…) Its insertion may be accounted for by the assumption that in Ibn Ezra's mind belief is closely associated to "intelligence" and may be an offshoot of the later'.

[16] *Book of Reasons*, p. 79; see also *Alcabitius: The introduction to Astrology*, p. 75.

[17] *Book of Reasons*, p. 75; see also *Alcabitius: The introduction to Astrology*, p. 65.

If the Moon is associated with Mercury in some aspect or in conjunction, and both [the Moon and Mercury] aspect the Ascendant degree, the man's soul will be perfect, his mind will overcome his desire, he will do everything honestly and lawfully.[18]

In this case, the condition is only partially fulfilled: Mercury aspects the Moon by a close trine and the Ascendant by an extended conjunction, but the Moon does not aspect the Ascendant.[19] However, as the Moon is strongly placed in its own rulership, Cancer, Abraham might have decided to rule this whole configuration as positive.

From another perspective, the 'love of knowledge and the company of wise men' could also be connected to the position of Jupiter, ruler of the Ascendant in Gemini (the domicile of Mercury) in the 7th house, and also from Venus, the planet of harmony, in the 1st house in Sagittarius.

In this case, Mercury is conjunct to the Ascendant.[20] Mercury is in joy in the Ascendant, being therefore considered to be very strong and prominent.[21]

3. *Et erit amator mulierum, sed caute et absconse.*
He will love women, but cautiously and in a hidden way.

As this comment is interpolated between the judgment of the native's mentality and the calculation of his life span, it can be taken as part of the interpretation for the first house. Loving women is part of the native's personality. The topic of wives and marriage is addressed in point 8. The love of women can be related to the prominent placement of Venus, the planet of women, in the Ascendant, and to the position of Mercury, ruler of the VII House of marriage, in the XII House of hidden things.

4. *Et vivet .lxviiii. annis et erit omnibus diebus vite sue sanus corpore, sed tamen habebit aliquantulum infirmitatis in stomacho ex calore.*
He will live for 69 years and will enjoy good health all the days of his life

---

[18] *Nativities*, p. 107.

[19] For an explanation of the aspects, see Ptolemy, *Tetrabiblos*, trans. F. E. Robbins (London: The Loeb Classical Library, 1940), pp. 73–75.

[20] Mercury is indicated as being in the eleventh house in 30° Scorpio. But since 30° of one sign = 0° of the next, Mercury is really in the Ascendant.

[21] Joy: when a planet is in a house that is particularly favourable to its nature, it is said to be in joy. The joys are: the first house to Mercury, the ninth to the Sun, the third to the Moon, the eleventh to Jupiter, the fifth to Venus, the twelfth to Saturn and sixth to Mars. See *Book of Reasons*, p. 217.

except that he will suffer a little illness in his stomach because of heat.

The illness in the stomach can be attributed to the square between the Moon and Mars. The Moon is a general significator of the body and its humours, and it rules over the stomach; Mars is a malefic planet, its maleficence deriving from the excess of heat. This topic will be resumed on point 12.

Longevity Calculation (abridged version)

This calculation relies on the determination of two important factors in the horoscope: the *hyleg* (or *haylaj*), the planet which signifies life, and the *alcocoden* (*al-kadhkhudah*), the planet which determines the length of life.

The *hyleg* is usually one of the luminaries – the Sun for diurnal horoscopes (that is, horoscopes with the Sun above the horizon); the Moon for nocturnal horoscopes (Sun below the horizon). In order to be considered the *hyleg* of the horoscope, the luminary has to be in favourable conditions (such as being placed in an appropriate sign and house, and not afflicted by other planets). If none of the luminaries is fit to be the *hyleg*, the lunation preceding birth, the Ascendant or the Part of Fortune will assume that role (the order differs in diurnal and nocturnal horoscopes). It is up to the astrologer to ponder all conditions and determine the appropriate *hyleg* for each horoscope.

Once the *hyleg* is defined, the astrologer can determine the *alcocoden*, the planet which regulates the number of years that the native will live. In order to be considered a suitable *alcocoden*, the planet has to fulfill two basic conditions: it has to be the ruler of one of dignities of the *hyleg* and it has to aspect it. The dignities are: rulership or domicile, exaltation, triplicity, term and face.[22]

To each planet is associated a certain number of years: greater, median or lesser.[23]

Years of the planets							
Years	Saturn	Jupiter	Mars	Sun	Venus	Mercury	Moon
Greater	57	79	66	120	82	76	108
Median	43.5	45.5	40.5	69.5	45	48	66.5
Lesser	30	12	15	19	8	20	25

The condition of the *alcocoden* in the nativity determines the number of years given to the native: a good condition signifies the planet's greater years, a median condition, the median years, and a bad condition the lesser years. (Other planets also contribute to this calculation, by adding or subtracting years or months; these conclusions are seen as possibilities, and require confirmation by other prediction methods.)

Fig. 4. Table III – Longevity Calculation (abridged version).[24]

---

[22] For a complete explanation about the dignities, see *Book of Reasons,* pp. 43–49; also Ptolemy, *Tetrabiblos,* pp. 78–113.

[23] There is also a fourth category, called 'maxima', which is not used to calculate human life span. The maximum years of the planets are: Saturn: 256; Jupiter: 426; Mars: 284; Sun: 1461; Venus: 1151; Mercury: 461; Moon: 520 years. They are usually used in mundane astrology. See *Book of Reasons,* pp. 71–83, 217–235, and also a reference on p. 133.

The calculation of longevity is also part of the interpretation of the first house. Its calculation is complex, as is extensively debated on *Sefer ha-Moladot*.[25] On this matter, Ibn Ezra quotes several authors, such as Ptolemy, Dorotheus and Masha'allah, who offer alternative versions of this calculation, and presents his own opinion on the matter. In spite of the variations, the basic concept of the method remains essentially the same.

Reconstructing Abraham's method
This reconstruction presents some of the possible calculations for this horoscope, according to Abraham's method.

Hypothesis 1
Being a diurnal chart,[26] the Sun would be the natural choice for the *hyleg*, but in this horoscope the Sun is in Scorpio in the twelfth house, with no strength other than face (which is weak, almost irrelevant) and placed in an unfortunate house.[27] If, in spite of its weakness, the Sun was taken as the *hyleg*, the *alcocoden* would be a planet that rules any of the dignities of the sign in which the Sun is placed AND also aspects the Sun to be 'active'.

As the Sun is in Scorpio, its rulers are (by order of importance):[28]

Throne/domicile – Mars
Exaltation – none
Triplicity – Venus, Moon and Mars
Term – Mercury (from 11° to 19°)
Face – Sun (from 10° to 20°)

---

[24] An explanation on the calculation of life span can be found in *Ibn Ezra on Nativities*, pp. 45–57, 109–121 and 257–260. For determining the *hyleg*, see *Book of Reasons*, pp. 237–239.

[25] For a more complete explanation, see the Table III in Figure 4 and the respective footnote.

[26] A horoscope is diurnal if it has the Sun above the horizon (that is, in in the twelfth, eleventh, tenth, ninth, eighth or seventh houses); it is nocturnal if the Sun is below the horizon (in the first, second, third, fourth, fifth or sixth houses). For the practical application of this concept to the calculation of longevity, see *Nativities*, pp. 109–111.

[27] For the meaning of 'face' (under the designation of 'decan') see *Alcabitius: The introduction to Astrology*, pp. 31, and *Nativities*, pp. 224 and p. 245.

[28] An explanation about rulership is provided in *Book of Reasons*, pp. 43–51 and in *Alcabitius: The introduction to Astrology*, pp. 29–31.

Of all these possible rulers only the Moon aspects the Sun by a major aspect, in this case a trine.[29] Therefore, the astrologer may have taken the Moon as *alcocoden*. Although the Moon is in its domicile in Cancer, which makes it very strong, it is also afflicted by the South Node and the square of Mars, and placed in an unfortunate house, the eighth; therefore, it is not in a condition to give its greater years, 108, but also not so bad that it would only give its lesser years, 25 – it would therefore give its median years, 66.

Hypothesis 2
If Abraham had rejected the Sun as the *hyleg* due to its debilities (in a feminine sign and in an unfortunate house, the twelfth), his second choice would have been the other luminary, the Moon (in spite of being in an unfortunate house, the eighth).

As the Moon is in Cancer, its dispositors are:
Throne/domicile – Moon
Exaltation – Jupiter
Triplicity – Venus, Moon and Mars
Term – Mercury (from 13°-19°)
Face – Mercury (from 10° to 20°)

In this case, the only suitable *alcocoden* would be Mercury, which rules the Moon's term and face (none of the other rulers aspects the Moon by a major aspect, therefore no other can be *alcocoden*).

If Mercury is the *alcocoden*, we must consider its conditions to determine how many years it can give. Mercury is at 23° Scorpio, where it is peregrine (that is, it has no dignity and no rulership there) and in the unfortunate twelfth house. Due to these conditions, it cannot give its greater years, 79, only its median years, 48.

Hypothesis 3
As neither the Sun nor the Moon seems to fulfil the conditions to be the *hyleg*, Abraham's third choice would have been the lunation preceding birth. In this case, the preceding lunation was the Full Moon, which occurred six days before, on the 23rd October, at 9:26 pm. At the moment of the lunation the Moon was above the horizon in the eleventh house, and in 6° of Taurus, strongly dignified by exaltation and triplicity. Its rulers are:

---

[29] This calculation includes the conjunction (0°) and the aspects of sextile (60°), the square (90°), the trine (120°) and the opposition (180°).

Throne/domicile – Venus
Exaltation – the Moon
Triplicity – Venus, Mars and Moon
Term – Venus (from 0° to 8°)
Face – Mercury (from 0° to 10°)

The main ruler for this degree of Taurus is obviously Venus, which is dignified by rulership, exaltation, triplicity and term, but Venus does not aspect the lunation degree by any major aspect, so it cannot be *alcocoden*.

The Moon, of the triplicity rulers for the lunation degree, aspects it by an extended sextile, so it is a possible *alcocoden*. As seen before, the Moon can only give its median years, 66, due to its conditions in this horoscope.

Another possibility is Mars, which is also lord of the triplicity and aspects the lunation degree by a weak opposition.[30] As Mars is debilitated by exile in the sign of Libra, but placed in a fortunate house, the eleventh, it can therefore be assumed that it gives its median years, 40 and a half.[31]

Hypothesis 4

If none of the above possibilities was considered a satisfactory *hyleg*, Abraham could have also chosen the Ascendant of the Part of Fortune, but this is a very unlikely choice in this horoscope (the Ascendant is only aspected by Mercury, by a weak conjunction, and the Part of Fortune has no major aspects).

It would therefore be reasonable to suggest that he selected the Moon as *alcocoden*, whether he has decided to follow the first or the third hypothesis.[32]

In any case, Abraham states that these calculations are 'only an approximation to the truth', and that the life span determined by the *alcocoden* has always to be confirmed by other calculations, namely a direction[33] suggesting death (for instance, the direction of the significator of life or the ruler of the Ascendant to a malefic planet – Mars or Saturn).[34]

---

[30] A 'weak aspect' is one that is almost outside the range of the planet's orbs, being therefore relatively irrelevant in the context of a particular horoscope.

[31] For further detail about the criteria of attribution of years, see *Nativities*, pp. 51–52 ('The years of the *Kadhkhudah*').

[32] These calculations may present minor variations, according to the criteria followed by each astrologer.

[33] For an explanation about directions see *Book of Reasons*, pp. 331–332.

[34] *Nativities*, pp. 115: 'know that everything I mentioned to you is only an approximation of the truth, for when we know the lifespan <according to the years

If the life span does not coincide with the directions, some adjustments are required:

> if at such a time [i.e., the lifespan according to the years of the ruler] <the direction> reaches a place of death, the native will die in that year, but if it reaches <a place of death> only after a number of years, the native will die only when the place of life reaches the place of death, although <the native> will live, the years that are added to the years of the ruler will be affected with many diseases.[35]

There are two primary directions suggesting death around the native's sixty-ninth year:

1. Sun to the opposition of the Moon (zodiacal direction; date: October 1205). The Moon is placed in the eighth in this horoscope (being therefore related to death), and the Sun is possibly the *hyleg*, and thus significator of life. (The corresponding converse direction occurred in December 1204.)
2. Sun to the square of Mars (zodiacal direction; date: August 1206); the Sun, possibly the *hyleg*, forms a square (a difficult aspect) with a malefic planet, Mars. This configuration is typically considered to be a sign of death. (Corresponding converse: April 1200, which is too soon to be taken in account, for the native would have been 64 years old).

5. *Et cum expleverit .xxv. annum et paulo amplius, accipiet uxorem convenientem de nobili prosapia et tunc incipiet fortuna sua meliorari in omnibus que facere voluerit; sed propter Iovem diliget mulierem.*
When he has reached his twentieth-fifth year or a little later, he will marry a suitable wife from a noble family and then his good fortune will start to improve in everything that he wants to do; [but] because of the position of Jupiter he will be affectionate towards his wife.

Though it mentions marriage, his judgment seems to be related mainly to the second house of property and resources, and only secondarily to the

---

of the ruler>, we need to direct the appropriate place of life to be directed', and also pp. 55–56 ('Coordinating the Years of the *Kadhkhudah* with the Direction of the *Haylaj*'). On another note, it is worth noticing that some astrologers used to perfect their calculations by taking into account the aspects of other planets to the *alcocoden*. According to their nature and their condition in the horoscope, the aspecting planets would add or subtract a few years or months to the basic lifespan.
[35] *Nativities*, p. 115.

seventh house of marriage. Abraham enumerates the conditions that suggest wealth:

> If Jupiter is in one of the cardines, or in the eleventh place, and is neither under the Sun's rays nor retrograde, it too indicates great good fortune, wealth, and money.[36]

This condition is only partially met in this horoscope, as Jupiter is in the seventh house (one of the cardines, that is, one of the angular houses), and free from the Sun's rays (that is, more than eight degrees away from the Sun, thus avoiding being obscured by the Sun's power).[37]

From another perspective, the placement of Jupiter in the seventh house also suggests benefice coming from marriage, as well as affection, dignity and correct behaviour between spouses.

Abraham states that this benefice will come to the native around his twenty-fifth year.

This precise timing is possibly related to one or more primary directions happening around that time. The main primary directions that can be related to this event are:

1. Ascendant to the terms term of Mercury (zodiacal direction; data: April 1155). The significator of the native enters the terms term of Mercury, the ruler of the seventh house: the native becomes inclined to marriage.
2. Part of Fortune to the square of Mercury (zodiacal direction; date: November 1154). Mercury is the ruler of the seventh house of marriage and the Part of Fortune promises wealth: benefits from marriage.
3. Venus to the conjunction of Saturn (mundane direction; date: June 1156). Saturn is the ruler of the second house in this horoscope, therefore signifying wealth, and Venus is a natural significator of women. Furthermore, these planets are also the 'parents' of the Part of Marriage (corresponding converse direction: July 1156).
4. Venus to the opposition of Jupiter (zodiacal direction; date: July 1156). Venus, the general significator of women, reaches Jupiter, a planet positioned in the seventh house in this horoscope (corresponding converse direction: April 1156).
5. Venus to the terms of Mercury (zodiacal direction; date: November 1156). Venus, the significator of women, enters the terms of Mercury, the ruler of the seventh house of marriage: another indication of proneness to marriage.

---

[36] *Nativities*, p. 123.

[37] The cardines or angular houses are the first, the fourth, the seventh and the tenth.

6. Sun to the term of Jupiter (zodiacal direction; date: April 1157). The Sun enters the term of Jupiter, the significator of wealth.
7. Moon to the term of Venus (zodiacal direction; date: June 1157). A general indication of events and activities related to women, as both Venus and the Moon have that signification.
8. Venus to the conjunction of Saturn (zodiacal direction; date: November 1157). Similar to direction number 3, but including latitude in the calculation (corresponding converse direction: August 1157).
9. Mercury to the conjunction of Venus (zodiacal direction; date: January 1158; mundane direction; date: February 1158). The ruler of the seventh house, Mercury, conjoins the significator of women, Venus (corresponding converse: same dates).

The inter-planetary directions (that is, the ones that involve planets or luminaries) are taken as clear indications of events, while the directions by term (which involve one planet and an abstract point of the horoscope) support and reinforce their significations.

From these directions it can be inferred that the period between 1155 and early 1158 is particularly favourable for both marriage and the acquisition of wealth (possibly through marriage). As the native was born in October 1135, he would between 19 and 22 years old in this period.[38]

Another possibility is the use of triplicities (in this particular case the rulers of the triplicity of the ascendant sign) to predict broader periods of life.

In this horoscope the Ascendant is Sagittarius, a fiery sign. The triplicity rulers for the Fire element are the Sun, Jupiter and Saturn – each one ruling over a third of the native's life. As he is supposed to live for 69 years, each third will be approximately 23 years. This means that the first third of his life (from birth to 23 years of age) is ruled by the Sun, the second (from 23 to 46 years of age) by Jupiter and the third (from 46 to death at 69 years of age) by Saturn. In this horoscope Jupiter is placed in the seventh house of marriage, so it is to be expected that marriage will be possible within that period; Jupiter is also a natural significator of wealth and abundance – hence the connection to marriage and riches.

---

[38] The author mentions the native's 25th year 'or a little later', but the main directions seem to point out to the period between his 19th and 22nd year. This discrepancy may be due to slight differences in the positions of the planets, since those presented in this text were calculated by computer. In primary directions, even the smallest difference may represent a discrepancy of several months or even years.

THE TRIPLICITIES
According to this technique, the astrologer divides the native's life into three thirds (which is only possible because he has previously calculated the native's life span) and attributes to each third one of the three rulers assigned to the Ascendant sign, according to its Element.
The sequence of rulers is as follows.

	Rulers for diurnal horoscopes[39]			Rulers for nocturnal horoscopes[40]		
Fire signs Aries, Leo, Sagittarius	Sun	Jupiter	Saturn	Jupiter	Sun	Saturn
Earth signs Taurus, Virgo, Capricorn	Venus	Moon	Mars	Moon	Venus	Mars
Air signs Gemini, Libra, Aquarius	Saturn	Mercury	Jupiter	Mercury	Saturn	Jupiter
Water signs Cancer, Scorpio, Pisces	Venus	Mars	Moon	Mars	Venus	Moon

The nature of the ruler for each period, as well as its condition in that particular horoscope, determines the general quality of the period (more precise predictions depend upon other techniques).[41]

Fig. 5. Table IV – The Triplicities.

It is also worth noticing that all these indications of wealth are reinforced by other positive conditions already patent in the horoscope, which will be discussed in point 9.

The remark about being affectionate towards his wife derives evidently from the position of the greater benefic, signifying not only wealth but also honour and joy, in the seventh house of marriage.

6. *Et aliquantulum fama blasphemabitur sed ei non nocebit.*

He will suffer a little from a malicious slander, but it will not harm him.
The 'malicious slander' can be related to Mercury in Scorpio in the twelfth

---

[39] As explained before, diurnal horoscopes are the ones with the Sun above the horizon.

[40] Horoscopes with the Sun below the horizon.

[41] Note that the rulers of the triplicities may also determine specific matters within each astrological house. For a more detailed explanation refer to *Nativities*, Appendix 8, pp. 465–474.

house. This configuration suggests sharp and aggressive speech (Mercury disposited by Mars) that happens in secret, hidden from the native (twelfth house of secret enemies and treasons). The ruler of the twelfth, Mars, is in Libra and placed in the eleventh house of friends and allies. This suggests a 'false friend', who is in fact a secret enemy. As Libra is in a human sign of the element Air, the harm may come from people (as opposed to harm caused by animals or by Nature) and mostly from words (which are spread through the air) – hence slander.

Also, Mars squares the Moon, which reinforces the notion of aggressiveness (even more in this horoscope, due to the rulership of Mars over the twelfth house). But Mars is weak in Libra (one of the signs of his exile), and that is probably the reason why Abraham states that this slander 'will not harm him'.

Following the order of the houses, this comment should be related to the third house of brothers. These are also related to Mars, as Ibn Ezra states that 'you should always observe Mars, because it signifies brothers in every nativity'.[42]

7. *Et habebit filios et filias, sed amplius de filiis.*
He will have sons and daughters, but more sons.

Clearly a judgment related to the fifth house of children. Ibn Ezra considers Jupiter as the main signifier for children:

> All the Ancients, including Ptolemy, agree that Jupiter signifies children, and that the children<'s fate> will be according to its [Jupiter's] position: If it is in a barren sign he [the native] will have few children, but the opposite applies if it is in one of the watery signs.[43]

As to the gender of the children, Ibn Ezra states:

> If it [Jupiter] is in a masculine quadrant, oriental to the Sun, and in a masculine sign, most of his children will be sons, but the opposite applies in the opposite case'.[44]

In this case Jupiter is placed in Gemini, a masculine sign and it is oriental to the Sun (that is, it is visible before the Sun in the west at sunrise). It is, however, in a feminine quadrant (which includes not only the seventh

---

[42] *Nativities*, p. 131.
[43] *Nativities*, p. 143–145.
[44] *Nativities*, p. 145.

house, where Jupiter is placed, but also the eighth and the ninth).

So, in this case, the planet fulfils only two out of the three necessary conditions: masculine sign and oriental to the Sun, but not masculine quadrant. Abraham judges accordingly, stating that the native's children will be both boys and girls, but mainly boys, due to the predominance of masculine factors.

8. *Et duas uxores habebit de quibus mortem videbit et bona illarum possidebit.*
He will have two wives, and will outlive them both and inherit their property.
This prediction relates to the seventh house of marriage and the eight house of inheritances. The notion of two wives may de attributed to the position of Venus (a significator of women) in a double-bodied sign.[45]

Quoting Enoch (i.e. Hermes), Abraham takes Venus as the main significator of wives, and lists the other significators for marriage:

> Always observe Venus, the Moon, the lot of women, the seventh place and its lord, and the lord of the seventh hour <after the nativity>, and determine the ruler over the aforementioned positions.[46]

The possibility of inheritance may be related to the position of Venus in Sagittarius, a sign ruled by Jupiter. As Jupiter is the ruler of the Ascendant in this horoscope – thus the significator of the native – it can be inferred that the native (Jupiter) will rule over his wives and will dispose of their belongings (Venus). This possibility is also patent in the primary directions discussed in point 5.

9. *Et quamdiu vixerit satis dives erit, quia in omnibus lucrabitur, et haec fortuna videtur ei contingere ex quo habebit .xx. annos.*
He will be sufficiently rich all his life, because he will make a profit out of everything; his good fortune will start from the age of twenty.
This sentence is the continuation of point 8. It is related not only to the second house, but again with the eighth house of inheritances and resources acquired from partnerships.

There are two main indications of wealth in this horoscope which, because they are part of the natal chart, will de 'active' throughout his life

---

[45] The double-bodied (also called mutable), signs are Gemini, Virgo, Sagittarius and Pisces. There is also a double-bodied sign in the cusp of the seventh house, Gemini, which reinforces the idea of several wives.
[46] *Nativities*, p. 157.

(that is, they do not depend upon directions to be 'activated').[47] They are the North Node in the second house (which suggests a general condition of abundance and growth) and Saturn, the ruler of the second house in this horoscope, conjunct to the cusp of the second, and in Capricorn, one of the signs it rules. The ruler of the house placed in the house (and strong) signifies that the matters of the house are relatively independent from the rest of the native's contingencies. In the case of resources, it also means that the native's wealth is easily improved, for it can generate itself – therefore, 'he will make a profit out of everything'.

These favourable conditions are supported by the angular position of Jupiter, which in itself promises wealth:

> If Jupiter is on one of the cardines, or in the eleventh place, and neither under the Sun's ray nor retrograde, it too indicates great good fortune, wealth and money.[48]

Jupiter's position in the horoscope can point out the age when this wealth will manifest:[49]

> If it [Jupiter] is in the first or tenth place, good fortune will come before the middle of his life; if it is in any of the other cardines, it will come after the middle of his life.[50]

In this case Jupiter is in the seventh house, a 'cardine' (that is, an angular house), but not in the first or the tenth; it is in the fourth, which suggests 'good fortune before the middle of his life'.[51] As Abraham has already

---

[47] See footnote 34 (*Nativities*, pp. 55–56).

[48] *Nativities*, p. 123.

[49] In this judgment the general condition of Jupiter in the horoscope has to be taken in account: 'His [the native's] wealth will be according to its [Jupiter's] power, depending on whether it [Jupiter] is the ruler of the nativity or the lord of the hour of the nativity or the lord of the Sun's house by day or of the Moon's house by night.', *Nativities*, p. 123.

[50] *Nativities*, p. 123.

[51] A planet located in one if the angular houses (the 'cardines') is considered to be strong, and therefore capable of manifesting its effects in full, as stated in *Book of Reasons*, p. 65. In another perspective, the seventh house pertains to the second quadrant (between the Mid-Heaven and the Descendant), which is related to youth; the third quadrant (between the Descendant and the lower Mid-Heaven) indicates the period 'when one approaches fifty years', as it is stated in *Nativities*,

determined that this native will live for sixty-nine years, we can now deduce that the middle of his life will be around his thirty-fifth year.

In sum, there will be two main favourable periods for the acquisition of wealth: around his twentieth year (possibly by marriage) and after his thirty-fifth year (by his own means, heritage or possibly a second marriage).

Note that Abraham's mention of 'the age of twenty' as the time when the native is likely to attain both riches and marriage may also be related to the directions listed in point 5. (For this topic, the most relevant directions are the ones related to Jupiter and Saturn, for the reasons already explained.)

10. *Et perget longa itinera pro sua lege inquirenda et quocumquei erit ab omnibus honerabitur et lucrabitur in omni mercatura, prout ipse voluerit.*

He will go on long journeys to find out more about his religion (pilgrimage), and wherever he goes he will be honoured by everybody and will make a profit in every business affair, just as he wants.

Long journeys and religion are two of the three topics pertaining to the ninth house (the third is wisdom, which was already addressed as part of the native's mentality, in the first-house interpretation).

The ruler of the ninth house in this horoscope is the Sun, which is placed in Scorpio, a fixed sign, suggesting a firm and constant attitude towards faith. The ninth is the house of joy of the Sun, and although in this horoscope the Sun is not in the ninth, it rules it, for the cusp is in Leo, further reinforcing the strength of the matters of this house.

Quoting Enoch (Hermes), Ibn Ezra states that the Sun ruling the ninth signifies 'prophetic dreams', thus connecting this house with the search for wisdom and truth (by several means).[52]

Furthermore, the Part of Fortune, meaning wealth and prosperity, is conjunct to the cusp of the IX House of journeys and religion, suggesting profits originating from the matters pertaining to this house: religion, knowledge and long journeys.[53]

---

pp. 366–367. The planet's location in the seventh – that is, in the end of the second quadrant – points out to the end of the youthful period, just before mid-life.

[52] For the connection of dreams to the ninth house, see *Nativities*, p. 171.

[53] *Nativities*, p. 129: 'You should also observe the Lot of Fortune, because its power is like the power of the second place, and any planet in aspect with it signifies money, according to it s[the planet's] nature'. In this case, the Part of Fortune is at 02°56' Leo, conjunct to the cusp of the ninth house (09°56' Leo).

11. *Et sicut etas augmentabitur ita et ipsa pecunia et ditior erit omnibus parentibus suis.*

As his age advances so will his wealth, and he will be more wealthy than all his ancestors.

This was already addressed in points 5, 8 and 9.

12. *Et morietur in terra sua ex forti valitudine stomachi cum par <vo> tussi.*

He will die in his own land from a strong stomach infection accompanied by a little cough.

In spite of his pilgrimages, the native will manage to return home to die – which in the Medieval period was considered the best possible choice. Dying at home allowed the native to be safe and, more importantly, it ensured the appropriate religious ceremonies would be performed.

The Moon, ruler of the VIII House of death, is placed in the VIII, so death will occur in its own 'land'. In this matter Ibn Ezra again quotes Enoch (Hermes):

> If the lord of the eighth place is in an auspicious position and is not aspected by a malefic, he will die in his bed from a disease that corresponds to the division of the signs, according to the method that assigns the head to Aries.[54]

As the Moon is strong in Cancer (the signs of her rulership), death should be relatively easy, but unfortunately the Moon is also aspecting Mars by square. This suggests that there will be some suffering involved – in this case an infection in the stomach, as was explained in point 4.

13. *Anni fortunati sunt ei 19, 21, 25, 28, 31, 37, 45, 49, 52, 61; anni infortunati sunt 23, 26, 35, 38, 47, 50, 59, 62; menses fortunati November, Februarius, Iulius; infortunati September, December; dies boni, dies Veneris, mali, dies Martis; nox bona, quam sequitur dies Martis; mala, quam sequitur sabbati.*

His lucky years are 19, 21, 25, 28, 31, 37, 45, 49, 52, 61; his unlucky years are 23, 26, 35, 38, 47, 50, 59, 62; his lucky months are November, February and July; his unlucky months are September and December; his good days are Fridays, his bad days, Tuesdays; his good night is the one that follows

---

Oddly enough, the Part is not represented in this horoscope (which also does not include any other parts or fixed stars).

[54] *Nativities*, p. 167. He repeats the concept a few lines further: 'If the lord of the eighth place aspects its place [the eighth place], the native will die in his own country', *Nativities*, p. 167.

Tuesday, his bad night, the one that follows the Sabbath.
It is unclear how Abraham arrived at these calculations.

14. *Et ille sit super omnia benedictus qui de hac re veritatem non ignorat.*
Let that man be blessed above all who is not unaware of the truth of this!

15. *Hanc nativitatem iudicavit Abraham Judeus Biterris.*
Abraham the Jew judged this nativity in Béziers.

Final comments
Though very concise, Abraham's interpretation encompasses the main points of the horoscope, the ones that would matter the most for a Medieval client: longevity, health, wealth, marriage, children (especially male children) and religion. Notably, there is no reference to the native's profession, but there are several mentions of wealth and profit, which suggests that the native is the son of a nobleman or of a rich merchant, and so he does not need to work for a living. The reference to slander emphasizes the importance of reputation in Medieval society. It is also interesting to notice that death is discussed in a quiet, almost casual manner. The foreknowledge of the time of death allowed the natives to organize their affairs and make all the necessary preparations to leave this world in peace – preferably in their own homes, surrounded by family and devoting their last moments to prayer.

The astrological interpretation reveals not only the client but also the astrologer. In this case, it is no surprise that the first topic addressed by Abraham is wisdom. He places the native's wisdom even before longevity – a choice that is perfectly consistent with his philosophical background. Judging from his profuse comments about resources and wealth, these concerns were very high in his priorities, and were often connected with other matters, such as marriage or religion. This can also be a hint to the patrons (possibly the parents of the native), to encourage their generosity towards the astrologer.

The lack of explanations to support the predictions, together with the absence of Parts and fixed stars in the horoscope, suggest some hastiness in the making of this text. The client was obviously not a king or a great prelate, to whom Abraham would have offered a more complete description.

Abraham's technique seems to be compatible to the astrological doctrine presented in Ibn Ezra's books. For instance, when interpreting the planets he combines three different perspectives:

- the planet's general meaning (i.e., mind for Mercury, abundance for Jupiter, obstacles for Saturn and so on);
- its specific position by house in a given horoscope (i.e., planets in the second house are related to wealth, in the fourth to children, in the tenth to honours, and so on);
- its specific rulership over the zodiac sign (or signs) in a certain house in a given horoscope (i.e, a planet ruling the cusp of the fourth house is related to origins, in the twelfth to enemies, and so on).

Following the astrological rules (as explained above) he is able to assemble these significations, according to the context of each topic, to produce a complex and detailed judgment. This combination of different layers of significance can also be found in Ibn Ezra's writings. The only significant difference is that this judgment is notably succinct, while Ibn Ezra offers copious explanations – and this can be attributed to the difference between writing a book and writing a judgment. Regardless, both texts present the marks of a seasoned astrologer and both follow compatible techniques, so it is reasonable to assume that they could have been written by the same author – Abraham Ibn Ezra.

Creation, Providence, and the Limits of Human Knowledge of the World: Mellin de Saint-Gelais and John Calvin on Astrology

Lindsay Starkey

Abstract: This paper examines John Calvin's (1509-1564) *Advertissement contre l'astrologie judiciaire* (1549) alongside Mellin de Saint-Gelais' (ca. 1490-1558) *Advertissement sur les iugemens d'astrologie à une studieuse damoyselle* (1546). Writing in the wake of the pressures that religious reformations had caused in France, both Calvin and Saint-Gelais assessed astrology based on assumptions from their common Christian heritage. For both, whether and how astrology was a valid pursuit depended on how God created the world, how he continued to relate to it through his providence, and the human ability to know the universe. Their diverse understandings of creation, providence, and the limits of human minds led them to judge the study of astrology differently. Analysing these themes in their works, this manuscript ultimately suggests that notions of creation, providence, and the boundaries of the human mind influenced sixteenth-century assessments of astrology and of the study of the universe more broadly as authors experienced the pressures of religious reformations on traditional Christian doctrines.

In a 1576 letter to Tycho Brahe (1546–1601), Charles de Dançay (1509–89), then the French ambassador to Denmark, included an anecdote about his own personal confrontation with John Calvin (1509–64) on the subject of astrology. Although more than thirty years had passed, Dançay explained that he had known Calvin during his time in Strasbourg from 1538–1541 and that they had disagreed over his – Dançay's – use of astrology. Dançay related that

> ... when he [Calvin] saw that I had consequently arrived at belief in such predictions that I boasted that I was able to make predictions concerning the adultery of women, some, which, I confess that it is true, had seldom had false

Lindsay Starkey, 'Creation, Providence, and the Limits of Human Knowledge of the World: Mellin de Saint-Gelais and John Calvin on Astrology', *Culture and Cosmos*, Vol. 18, no. 2, Spring/Summer 2014, pp. 41-70.
www.CultureAndCosmos.org

results, he rebuked me through his authority and heavy judgment, and I was
threatened severely, if I did not give it up. [1]

Dançay claimed to Brahe that he had not dared to touch such things again
after Calvin's stern warning.[2]

Dançay's letter provides a colourful and perhaps expected story of John
Calvin's response to astrology. Both scholarly and popular literature tend
to depict Calvin almost exclusively as a theologian, and there was a long
tradition of Christian theological critiques of astrology by the sixteenth
century.[3] Calvin was also known to have had a quick temper.[4] Yet his
irascible personality and his position as a religious reformer provide only a
superficial explanation for Calvin's reaction to Dançay's astrological
predictions. Almost certainly receiving formal, university training in
astrology during his studies at the University of Paris,[5] he also commented

---

[1] 'Is enim, cum vidisset, me eo credulitatis in hisce praedictionibus pervenisse, ut
de adulteriis etiam mulierum hinc nonnulla, quae, ut fatear quod verum est, raro
falsos habuerint eventus, praedicere me posse iactarem, pro sua autoritate et gravi
iudicio me increpavit, et male mihi minatus est, si ista non relinquerem.' Charles
de Dançay, 'Carolus Danzaeus ad Tychonem 17 Nov. 1576', in *Tychonis Brahe
Dani Opera Omnia*, ed. J. L. E. Dreyer (Copenhagen: Libraria Gyldendaliana,
1924), 7:41.
[2] Dançay, 'Carolus Danzaeus ad Tychonem 17 Nov. 1576', 7:41.
[3] On this tradition and its impact in the Renaissance, see Eugenio Garin, *Astrology
in the Renaissance: The Zodaic of Life,* trans. Carolyn Jackson, June Allen,
Eugenio Garin, and Clare Robertson (London: Routledge, 1983); and Anthony
Grafton, *Cardano's Cosmos: The Worlds and Works of a Renaissance Astrologer*
(Cambridge, MA: Harvard University Press, 1999). On the impact of religious
reformations on this theological tradition, see the articles in *Divination et
controverse religieuse en France au XVIe siècle*, in *Cahiers V. L. Saulnier 4: Actes
du Colloque organisé par l'Université de Paris-Sorbonne le 13 mars 1986 par le
Centre V. L. Saulnier* (Paris, 1987); and Robin B. Barnes, 'Astrology and the
Confessions in the Empire, c. 1550-1620', in *Confessionalization in Europe, 1550-
1700: Essays in Honor and Memory of Bodo Nischan,* ed. John M. Headley, Hans
J. Hillerbrand, and Anthony J. Papalas (Aldershot: Ashgate, 2004), pp. 131–153.
[4] One of Calvin's recent biographers, Bruce Gordon, describes Calvin as 'ruthless'
and as 'an outstanding hater'. See his *Calvin* (New Haven: Yale University Press,
2009), pp. vii–xi.
[5] On Calvin's childhood and education, see Abel Lefranc, *La jeunesse de Calvin*
(Paris: Fisbacher, 1888); Quirinus Breen, *John Calvin: A Study in French
Humanism,* 2nd edition (Hamden, CT: Archon Books, 1968); and Alexandre

on and referenced it in many of his works.[6] Calvin's educational background and his continued discussion of it suggest that his chastisement of Dançay was more than just a rehashing of theologians' opinions. Instead, it suggests that this chastisement came from Calvin's own conception of the content and the limits of astrology developed through his education in Paris and through his continued engagement with astrological texts.

Calvin's most sustained examination of the topic appeared approximately ten years after he rebuked Dançay in Strasbourg. In 1549, he explored the legitimacy and the scope of astrology at length in a treatise entitled, *Advertissement contre l'astrologie judiciaire*.[7] An investigation of this treatise and the text that likely provoked its composition permits the exploration of his conception of astrology as well as the questions that surrounded astrology's study in sixteenth-century Europe. Though resonating with medieval and Renaissance discussions of the subject, Calvin's treatise was likely a more immediate reply to one that Mellin de Saint-Gelais (ca. 1490–1558), a court poet to Francis I and Henry II of France, had written in 1546, *Advertissement sur les iugemens d'astrologie*

Ganoczy, *The Young Calvin*, trans. David Foxgrover and Wade Provo (Philadelphia, PA: Westminster, 1987).

[6] For example, Calvin spoke about astrology at length toward the end of his life in his 1563 commentary on Jeremiah as he commented on Jeremiah 10:2. See *Corpus Reformatorum Ioannis Calvini Opera quae supersunt omnia*, Vol. 38, ed. Guilielmus Baum, Eduardus Cunitz, and Eduardus Reuss (Braunschweig, C. A. Schwetschke and Sons, 1888), Cols. 57–62. On the importance of astrology and the world more generally to Calvin's theology, see Susan E. Schreiner, 'Creation and Providence,' in *The Calvin Handbook*, ed. Herman J. Selderhuis (Grand Rapids, MI: William B. Eerdmans Publishing, 2009), pp. 267–274; and Randall C. Zachman, 'The Beauty and Terror of the Universe: John Calvin and Blaise Pascal,' in *Reconsidering John Calvin* (Cambridge: Cambridge University Press, 2012), pp. 6–34.

[7] On the treaty's structure as well as its relationship to Calvin's discussions of astrology in his biblical commentaries, see Christine McCall Probes, 'Calvin on Astrology', *Westminster Theological Journal* 37, no. 1 (1974): pp. 24–33. For a comparison of Calvin's assessment of astrology in this treatise to the judgments of Martin Luther, Philipp Melanchthon, and François Rabelais, see John Lewis, 'Pronostications et propaganda évangélique,' in R. Aulotte, *Divination et controverse religieuse en France au XVIe siècle, Cahiers 50* (1987): pp. 73–83. For this treatise's most recent English translation, see Mary Potter, 'A Warning against Judicial Astrology,' *Calvin Theological Journal* 18 (1983): pp. 157–189.

à une studieuse damoyselle.[8] Saint-Gelais is better known today for his court poetry and as a court rival to Pierre de Ronsard (1524–85) than as a writer of treatises on subjects such as astrology.[9] His text took the form of a letter addressed to a young woman. Professing to answer her query about the acceptability of astrology, Saint-Gelais argued that its study was permissible. Calvin assessed this study differently. Dividing astrology into two types, he claimed that the practice of illicit astrology was increasing among his contemporaries and that he wrote for the simple and the unlearned, who he feared would fall into its trap.[10] Writing in the wake of religious reformations, both Saint-Gelais and Calvin re-examined many of the topics that had long been a part of earlier judgments of astrology. A comparison of their works therefore allows us to begin to investigate the impact that these religious reformations had on assessments of astrology. The modern scholar, Olivier Millet, has argued for the importance of viewing Calvin's *Advertissement* within the context of earlier debates about astrology's legitimacy. Though Millet provides an in-depth examination of Renaissance discussions of the subject, particularly as they occurred on the Italian peninsula and in France, he does not investigate the differences between Calvin and Saint-Gelais' notions of astrology in detail nor how the tensions that calls for religious reform had created in France by the 1540s might have influenced their views of it.[11] A comparison of Saint-Gelais and Calvin's assessments of astrology reveals that assumptions from their shared Christian heritage were at the centre of their

---

[8] The modern editor of Calvin's *Advertissement contre l'astrologie judiciare*, Oliver Millet, argues for the likelihood that Calvin wrote to counter Saint-Gelais' arguments. See Millet, 'Introduction', in Jean Calvin, *Advertissement contre l'astrologie judiciaire*, ed. Oliver Millet (Geneva: Librairie Droz, 1985), pp. 22–28.

[9] Henri-Joseph Molinier, *Mellin de Saint-Gelays: Étude sur sa vie et sur ses oeuvres*, Reprint (Geneva: Slatkine, 1968).

[10] Calvin, *Advertissement contre l'astrologie judiciaire*, pp. 51–53.

[11] See Millet, 'Introduction'. The comparison between Saint-Gelais' notion of astrology and Calvin's appears on page 28. Millet argues that personal temperament and a different religious outlook – Calvin, the militant, confessionalizing evangelical, and Saint-Gelais, the non-specialist, lay person – led them to argue for different conceptions of astrology. Assuming that Saint-Gelais and Calvin's religious outlooks determined their judgments of astrology, Millet does not explore how religious reformations influenced the very questions that sixteenth-century authors might have asked about astrology regardless of their personal religious views.

different views of astrology. As Christians, both Calvin and Saint-Gelais professed the belief that God had created the world. This assumption involved much more than just the acknowledgement of God's status as creator. For both men, the presupposition of the universe's creation included the belief that the ways in which the world currently operated had something to do with how God had designed it. It also incorporated the belief that God had fashioned people a certain way during the creation process. What effects celestial bodies had on the terrestrial realm and whether people could know these effects depended on how God had designed the world and the human beings who tried to gain knowledge of them. Two further presuppositions also figured into their judgments of astrology. For Calvin and Saint-Gelais, God continued to administer the world in the present day through his providence. For them, the Fall had also damaged prelapsarian human capacities to know the world. God's continued providential control over the universe and the postlapsarian human ability to gain knowledge of it also affected whether studying astrology was a valid pursuit. Though sharing these assumptions, they had different views of how God had made the world, how he continued to relate to it, and the contemporary human capacity for knowledge of it. Their diverse understandings of creation, providence, and the limits of human minds ultimately led them to assess the study of astrology differently. A comparison of Calvin and Saint-Gelais' texts also suggests that the very understandings of creation, providence, and the human knowledge of the world were in question for sixteenth-century authors as they experienced the pressures of religious reformations. Many previous authors on astrology shared Saint-Gelais and Calvin's assumptions of God's creation of and continued providential control over the world. The latter assumption especially figured into earlier assessments of astrology. Prompted to reconsider traditional Christian doctrines in the wake of religious reform movements, Saint-Gelais and Calvin differed from these authors in their particular focus on how God's providence related to celestial influences and on the boundaries of human knowledge of God's creation. Analysing these themes in their works sheds light on how gradually shifting notions of creation, providence, and the limitations of human minds influenced the sixteenth-century assessment and the study of astrology and of the universe more broadly.

Astrology and Its Assessment before Saint-Gelais and Calvin
Much like the vast majority of their contemporaries, both Saint-Gelais and Calvin held that celestial bodies had some influence on the terrestrial

realm.[12] They both called the study of these influences, 'astrology'. They advanced different notions of the particular effects that the heavens had on the earth and of how people could study the heavenly bodies to predict these effects. As they assessed these influences and their study, Saint-Gelais and Calvin drew on the long tradition of writing about astrology and the study of celestial bodies more generally. Examining this tradition is essential for understanding how significant Calvin and Saint-Gelais' notions of creation, providence, and the boundaries of human knowledge were to their judgments of astrology.

Writing on astrology in the sixteenth century, Saint-Gelais and Calvin inherited a classification system for knowledge of celestial bodies and their effects that had developed from the ancient period and had become relatively stable by at least the thirteenth century. Ptolemy (ca. 90–168) had divided the study of celestial bodies into two parts in his *Tetrabiblos*. Calling both of them astronomy, he argued that the first was the one 'whereby we apprehend the aspects of the movements of the sun, moon, and stars in relation to each other and to the earth', and that the second was the one 'in which by means of the natural character of these aspects themselves we investigate the changes which they bring about in that which they surround'.[13]

Saint-Gelais and Calvin along with other sixteenth-century authors made a similar division, though they tended to call the first kind of inquiry, astronomy, and the second, astrology. They equated astronomy with the study of the positions and motions of celestial bodies. They then defined astrology as the study of celestial influences based on knowledge of these positions and motions. Though this division of the study of the stars was common, the consistent usage of the terms, 'astronomy' and 'astrology' was not. Authors often used the terms, 'astrology' or 'astronomy', to refer to both types of study much as Ptolemy had done.[14] They also further

---

[12] On the supposed influences of celestial bodies and their relationship to astrology see, John. D. North, 'Celestial Influences – the major premiss of astrology', in *'Astrologi hallucinati': Stars and the End of the World in Luther's Time*, ed. Paulo Zambelli (Berlin: Walter de Gruyter, 1986), pp. 45–100.

[13] Ptolemy, *Tetrabiblios*, I.1. The translation comes from the 1940 Loeb edition, trans. F. Boll and Emilie Boer, p. 3.

[14] On the various knowledge categories involved in the study of celestial influences as well as the forms of representation of that knowledge by the sixteenth century, see Robert S. Westman, *The Copernican Question:*

classified either 'astronomy' or 'astrology' to delineate different types of knowledge of celestial bodies and their influences. Calvin is a good example. He did not differentiate consistently between 'astronomy' and 'astrology' Instead, he wrote of a 'natural astrology' that explored both the positions and motions of celestial bodies and their natural effects on the terrestrial realm and a 'judicial astrology' that attempted to predict the future course of human actions or events on the earth based on the knowledge of celestial bodies.[15] As Calvin explained:

> Therefore, it must be admitted that there is some correspondence between the stars and planets and the dispositions of human bodies. All of this, as I have said, is included under natural astrology. But those shameless deceivers, who want, under the shadow of this art, to go beyond its proper limits, have contrived another type of astrology, which they call "judicial." It consists of two principle articles: first, it is not only the knowledge (*savoir*) of the nature and complexions of people but also their fortunes, as they call them, and all that they will either do or suffer in their lives; second, it is to know what results all their enterprises will have in their dealings with each other, and in general the whole state of the world.[16]

In addition to inheriting a knowledge classification system, Saint-Gelais and Calvin also drew on a long-standing debate among Christians over whether the study of celestial influences was proper for Christians to undertake. Many Church Fathers, including Augustine of Hippo (354–430), had condemned the notion that celestial influences determined the course of individual human lives based on its seeming inconsistency with

---

Prognostication, Skepticism, and Celestial Order (Berkeley, CA: University of California Press, 2011), pp. 29–40.

[15] Calvin, *Advertissement contre l'astrologie judiciaire*, pp. 51–56.

[16] 'Ainsi il faut bien confesser qu'il y a quelque convenance entre les estoilles ou planettes et la disposition des corps humains. Tout cecy, comme j'ay dit, est comprins souz l'astrologie naturelle. Mais les affronteur qui ont voulu, souz ombre de l'art, passer plus outre en ont controuvé une autre espece qu'ilz ont nommée judiciaire, laquelle gist en deux articles principaux: c'est la savoir non seulement la nature et complexion des hommes, mais aussi toutes leurs avantures, qu'on appelle, et tout, ce qu'ilz doyvent ou faire ou souffrir en leur vie; secondement, quelles yssues doyvent avoir les entreprises qu'ilz font, traffiquans les uns avec les autres, et en général de tout l'estat du monde.' Calvin, *Advertissement contre l'astrologie judiciaire*, p. 57. My translations of Calvin's *Advertissement contre l'astrologie judiciaire* build on the translation of Potter, 'A Warning against Judicial Astrology'. For this passage, see Potter, p. 168.

the Christian doctrines of the human free will and divine providence.
Though Augustine did claim that celestial bodies influenced the earth, he
argued against the notion that the positions of celestial bodies at one's birth
caused one's future actions.[17] If the stars determined future behaviour, then
people would not be responsible for their actions nor would God have
actual control over his creation. Going a step further, Augustine asserted
that anyone who had successfully predicted a person's future actions based
on the casting of horoscopes did so only through the aid of demons.[18]
Isidore of Seville (ca. 560–636) offered a clarification of this distinction in
his *Etymologies*, explaining that the study of natural astrology investigated
the effects of the sun, moon, and stars on the earth, whereas the study of
superstitious astrology tried to tell the future based on the stars as well as
to predict how the stars would affect the character of individual people – a
precursor of the separation Calvin later made between natural and judicial
astrology.[19] The study of celestial influences on the earth continued to
grow especially from the twelfth century. Drawing on Muslim
commentaries on ancient authors as well as on Muslim works on astrology,
medieval scholars such as Albert the Great (ca. 1200–80) and Thomas
Aquinas (1225–74) allowed a place for the study of celestial influences in
European scholarship and society while building on the distinctions
Augustine and Isidore had made. In the second book of his *Summa
Theologiae*, Aquinas argued that the heavens did influence the earthly
realm, including human bodies. He explained that astrologers often made
accurate predictions about future human actions based on their studies of
celestial bodies because people usually followed their passions, which
were tied directly to their bodies.[20] However, he also claimed that the
celestial bodies could not directly affect the soul, meaning that human
beings ultimately retained their free will and therefore responsibility for
their actions.[21] This support for the notion that the study of celestial
influences could allow people to predict their effects on individual human
bodies encouraged some European universities to make such study an
integral part of their offerings by the fourteenth century. For example, the
universities of Padua and Bologna taught astrology as a regular part of

---

[17] Augustine, *De civ. D. Patrologia Latina* edition, IV.11 and V.1.
[18] Augustine, *De civ. D.* V.1–7; *De gen. ad lit.* II. 35–37; Conf. IV.4 and VII.8–10.
[19] Isidore of Seville, *Etymologiam sive originum*, ed. W. M. Lindsay (Oxford:
Oxford University Press, 1985), 1:148.
[20] Thomas Aquinas, *Summa theologiae*, Leonine Edition, I. Q 115 Art 3–4.
[21] Aquinas, *Summa theologiae*, I. Q 115 Art. 4.

their curricula from the early fifteenth century.[22] Degree candidates could encounter it during three different periods of their university training: during lectures on the *quadrivium* as they applied mathematics to the motions of the planets and stars and learned how to predict their practical effects, during natural philosophy lectures as students learned Aristotle's theories of celestial motion and of the relationship between the celestial and terrestrial realms and incorporated these theories into predications of the effects of planets and stars on the earth, and during medical lectures as candidates for a medical degree learned how the planets and stars could affect their future patients' bodies.[23] Though many scholars discussed the boundaries between celestial influences, God's providence, and the human free will and though a few such as Nicole Oresme (ca. 1320–82) acknowledged only the reality of solar and lunar influences but not predictions based upon them, astrology had found institutional support in universities by the late middle ages.[24] Men with university training often found employment at noble courts, providing both sought-after advice as well as entertainment for nobles and their courtiers.[25] The ways in which authors and printers employed the printing press also gave astrology a much wider audience than those with university training. Prognostications and horoscopes poured off European printing presses from the mid-fifteenth century, making astrology an aspect of everyday life.[26] By the end of the fifteenth century and the beginning of the sixteenth century, the study of celestial influences had become popular in Europe. Discussions of celestial influences occurred in universities, noble courts, and among

---

[22] H. Darrel Rutkin, 'Astrology', in *The Cambridge History of Science*, Vol. 3: *Early Modern Science*, ed. Katharine Park and Lorraine Dastone (Cambridge: Cambridge University Press, 2006), pp. 542–547.

[23] H. Darrel Rutkin, 'Astrology, Natural Philosophy, and the History of Science, c. 1250–1700: Studies toward an Interpretation of Giovanni Pico della Mirandola's *Disputationes adversus astrologiam divinatricem*' (PhD Dissertation, Indiana University, 2002), pp. 36–160.

[24] G.W. Coopland, *Nicole Oresme and the Astrologers: A Study of his* Livre de Divinacions (Liverpool: Liverpool University Press, 1952), p. 57.

[25] Michael H. Shank, 'Academic Consulting in Late Medieval Vienna: The Case of Astrology', in *Texts and Contexts in Ancient and Medieval Science: Studies on the Occasion of John E. Murdoch's Seventieth Birthday*, ed. Michael McVaugh and Edith Sylla (Leiden: Brill, 1997), pp. 245–270.

[26] Bernard Capp, *English Almanacs, 1500-1800: Astrology and the Popular Press* (Ithaca, NY: Cornell University Press, 1976). See also Grafton, *Cardano's Cosmos*, pp. 71–108; and Westman, *The Copernican Question*, pp. 25–28.

people, who read, heard, or saw the widely circulated prognostications. A few late fifteenth-century authors began to reassess the legitimacy of studying celestial influences.[27] Robert S. Westman has recently singled out the works of Giovanni Pico della Mirandola (1463–94) and Lucio Bellanti (n.d.–1499) as particularly influential works in this reassessment.[28] Pico's *Disputationes adversus astrologiam divinatricem* (written 1493–1494, printed 1496) and Bellanti's *De astrologica veritate liber quaestionum* (1498) attempted to explain what astrology was and how celestial bodies affected the terrestrial realm. Even though Pico and Bellanti described astrology in similar ways – as the study of the effects that the positions and the motions of celestial bodies had on the earth – they judged this study much differently. Diverse views of how God had made the world as well as of how human beings could know about it were the fundamental divergence between Pico and Bellanti's assessments of astrology. For Pico, God had made the universe and human beings in such a way that they could not accurately predict how celestial bodies such as the planets and stars would affect the earth and individual human beings. As Pico explained in the second book of the *Disputationes,* God had designed the world so that it would reveal some of his attributes. Pico argued that the heavens in particular displayed these attributes especially to those who studied natural philosophy as they carefully observed the heavens to discern these invisible attributes through their visible motions.[29] In contrast, however, 'the astrologer, invited to view the heavens, turned his eyes away from the heavens to the earth'.[30] Bellanti, on the other hand, claimed that God had fashioned the world so that celestial bodies did have predictable effects on the earth. For him, astrology allowed human beings to gain knowledge of these predictable effects.[31] Pico and Bellanti's

---

[27] Rutkin, 'Astrology', pp. 547–548.

[28] Westman, *The Copernican Question*, p. 172.

[29] Giovanni Pico della Mirandola, *Disputationes adversus astrologiam divinatricem*, 2 vols, ed. Eugenio Garin (Florence: Vallecchi Editore, 1946–1952), 1:116.

[30] 'Astrologus, si est aliquid, caelum invite videt et de caelo ad terram oculos deflectit; solus philosophus ex visibilibus invisibilia speculatur.' Pico della Mirandola, *Disputationes*, 1:126.

[31] Lucio Bellanti, *De astrologica veritate* (Florence: Gherardus de Haerlem, 1498), sig. air–aiiir. On this debate, see Remo Catani, 'The Polemics on Astrology, 1489-1525', *Culture and Cosmos* 3, no. 2 (1999): pp. 16–30; and Steven Vanden Broecke, *The Limits of Influence: Pico, Louvain, and the Crisis of Renaissance*

judgments of astrology along with the particular arguments and evidence that they used to make their judgments resonate with those in Calvin and Saint-Gelais' works, but a further circumstance provides the immediate context in which both Saint-Gelais and Calvin wrote. Though Bellanti's answer to Pico had stirred up much debate about the heavens' influence, sixteenth-century religious reformations added another layer of complexity to these debates. Religious reformations complicated them from the perspectives of theology, scriptural exegesis, and politics. As sixteenth-century Europeans discussed how Christian communities should understand and practice their Christianity and what role Scripture should play in these judgments, political unrest and persecution occurred on a large scale. This re-examination of basic Christian doctrines fed further debate about the study of celestial influences and their ability to signal the apocalyptic end of the world. The study of celestial influences could also provide evidence of a group's supposed unchristian behaviour if they were judged to trust the stars' influence more than God's providence – a potential justification for attacking those considered to be unchristian.[32] France in particular experienced the pressures of religious upheaval on assessments of astrology.[33] In addition to including discussions of celestial influences in training at the University of Paris and the publication of both cheap prognostications and learned treatises on the subject, the arrival at court of Catherine de 'Medici (1519–89) in 1533 to marry the king's son brought a person with particular interest in celestial influences to a centre of power in France. Endorsement of astrology was not universal there, however. In 1540, Giovanni Ferrerio (1502–79) published his *De vera cometae significatione contra Astrologorum omnium vanitatum* that set out Pico's arguments against the study of most celestial influences and placed

---

Astrology (Leiden: Brill, 2003). On the influence of Pico's *Disputationes adversus astrologiam divinatricem* on subsequent assessments of astrology and the teaching of astrology in European universities more generally, see Rutkin, 'Astrology, Natural Philosophy, and the History of Science'.

[32] On the impact of religious reformations on judgments of astrology, see the articles in '*Astrologi hallucinati:*' *Stars and the End of the World in Luther's Time*, ed. Paulo Zambelli (Berlin: Walter de Gruyter, 1986); Luc Racaut, 'A Protestant or Catholic Superstition? Astrology and Eschatology during the French Wars of Religion', in *Religion and Superstition in Reformation Europe*, ed. Helen Parish and William G. Naphy (Manchester: Manchester University Press, 2002): pp. 154–169; and Westman, *The Copernican Question*, pp. 109–140.

[33] On astrology in France in the 1530s and 1540s, see Millet, 'Introduction', pp. 17–22.

this critique in the context of the contemporary religious and political
upheaval that engulfed France.[34] By the time Saint-Gelais and Calvin wrote
their treatises, the nature of celestial influences and whether and how
people could study them had become an especially debated topic in France.
Saint-Gelais and Calvin's assessments of astrology addressed those
surrounded by these debates. Though they drew on a long tradition of the
study of the heavens and their effects on the earth, the more recent
pressures of religious reformations formed the immediate context in which
Saint-Gelais and Calvin wrote. This pressure allowed Saint-Gelais and
Calvin to re-examine topics that had long been included in assessments of
astrology. In this context, Saint-Gelais and Calvin considered some of the
assumptions surrounding God, his creation and administration of the
world, and the human knowledge of it in order to judge how celestial
bodies influenced the earth and whether and how people could truly
discover these influences. Their views of creation, providence, and the
limits of human knowledge were at the centre of their judgments of the
legitimacy and the scope of astrology.

**Mellin de Saint-Gelais and his *Advertissement sur les iugemens
d'astrologie à une studieuse damoyselle* (1546)**
Written in the midst of the discussions of religious reform in France, Saint-
Gelais' *Advertissement sur les iugemens d'astrologie à une studieuse
damoyselle* reveals his concern for creation, providence, and the
boundaries of human knowledge of the world. No one attributed this
treatise to him until the nineteenth century. The original text itself did not
have a listed author. Its title page explained only that Jean de Tournes had
printed it in Lyon.[35] In 1866, M. Eusèbe Castaigne definitively attributed
the text to Saint-Gelais, pointing out that the sonnet that followed the text's
title page also appeared in a 1574 edition of his poetic works.[36] Though
better known for his poetry, Saint-Gelais argued strongly for the
permissibility of the study of astrology in this treatise. For him, the ways in
which God had fashioned the world and human beings during the creation

---

[34] On these discussions, see Mark Greengrass, *The French Reformation* (Oxford:
Oxford University Press, 1987), pp. 21–41.
[35] Mellin de Saint-Gelais, *Advertissement sur les iugemens d'astrologie à une
studieuse damoyselle* (Lyon: Jean de Tournes, 1546), a1r.
[36] See the introduction to *Advertissement sur les iugemens d'astrologie: Nouvelle
édition*, ed. M. Eusèbe Castaigne (Angoulême: F. Goumard,1866). See Saint-
Gelais, *Advertissement*, sig. a1v, for the sonnet in the original edition.

process meant that people could investigate the positions and motions of celestial bodies to predict their future effects on the terrestrial realm, including on people.

Saint-Gelais defined astrology as the 'science, which teaches one to judge things to come through the influence of celestial bodies'.[37] Stating that such study was permissible and even encouraged, he adopted a number of different argumentation strategies to support this argument. Many of these strategies had appeared in earlier endorsements of astrological predictions such as that of Lucio Bellanti. Saint-Gelais began his work with a chronological discussion of authorities who had endorsed its study, focusing especially on the ancient Egyptians, Babylonians, Persians, Greeks, and Romans.[38] Throughout this chronological discussion of authorities, Saint-Gelais also adopted a second strategy to prove the validity of astrology as he insisted repeatedly that firm epistemological foundations undergirded it. Again echoing earlier arguments, he pointed to astrology's basis in mathematics and claimed that it 'is founded by very obvious demonstrations, which cannot be denied'.[39] Insisting on astrology's basis in mathematics and demonstrative dialectic, Saint-Gelais then adopted a third argumentation strategy as he argued for a separation between this mathematically based astrology and the predictions that untrained people made. Saint-Gelais accused critics of astrology such as Giovanni Pico della Mirandola of ignoring this separation as they focused only on the unsupported, backward predictions of the vulgar and ignored the mathematical, firm basis of actual astrology.[40] Drawing on traditional defences, such as the one Ptolemy had offered to support the notion that celestial bodies did have effects on the earth that allowed astrologers to

---

[37] 'Si donque en toutes choses il y ha si differentes opinions, & mesmement en celles, qui se touchent & voyent à l'oeil: ce n'est merveille, si la science qui enseigne par l'influence des corps celestes a iuger des choses à venir, treuue divers iugemens de soy...', Saint-Gelais, *Advertissement*, sigs. a4v. (Unless otherwise noted, all translations are my own).

[38] 'So great is that utility, or rather the necessity of good and true opinions.' 'Grande est doncques l'utilité, ou plustost la neceßité des bonnes & vrayes opinions.', Saint-Gelais, *Advertissement*, sig. a3r. The discussion of authorities appears on sigs. a3r–a8v. Millet notes Saint-Gelais' reliance on authorities to support his argument for astrology's legitimacy from 22–28.

[39] 'Car comme astrologie est fondee sur demonstrations si evidentes, qu'on ne les peult nyer...', Saint-Gelais, *Advertissement*, sigs. a4v–a5r. (Unless otherwise noted, all translations are my own).

[40] Saint-Gelais, *Advertissement*, sigs. a8v–b1r.

predict the future,[41] Saint-Gelais' final strategy to show the legitimacy of such study relied on his understandings of creation, providence, and the human capacity for knowledge of the world. As he described the work of astrologers, he argued that 'the prediction of future things is not the main goal of true astrologers, but is only an accessory knowledge (*savoir*)'.[42] He explained that these true astrologers only gained the knowledge to predict the future because of their more general focus on the qualities and the natures of all things and on the correspondence and alliance between the celestial and the terrestrial realms.[43] For Saint-Gelais, God was responsible for this accessory knowledge, however, because he had designed the world so that celestial movements did influence the terrestrial realm, allowing true astrologers to make predictions involving individual human lives based on them. He judged this study thus: 'This is the divine source of this part of astrology, which is not acquired through vain and exterior observations, points, words, and images, but through the most enormous elevation of the spirit, and through the conduit of continual labor, illustrating the express favor and grace of God'.[44] Since God created the motions and effects of the planets and stars, the source of astrological knowledge was ultimately divine.

Saint-Gelais also claimed that God had fashioned people with intelligence and the knowledge that he had created the world so that people could use astrology to predict future occurrences. The significance of this connection between God's creation of the world and people as well as the human ability to gain knowledge of the world through astrology becomes clearest as Saint-Gelais wrote about how Christians in particular should relate to astrology. Rather than just stating that it was permissible for them to study it, he insisted that its study was a Christian obligation because God's designing of both the world and human beings gave them a special

---

[41] Saint-Gelais even called Ptolemy 'Philosophe excellent', Saint-Gelais, *Advertissement*, sig. a5v.

[42] 'La prediction des choses futures n'est point le but principal des vrayes Astrologues, mais est seulement un scavoir accessoire...', Saint-Gelais, *Advertissement*, sig. c1v.

[43] Saint-Gelais, *Advertissement*, sig. c1v.

[44] 'Voila la divine source de ceste part d'Astrologie qui ne s'acquiert par vaines, et exterieures observations, poinctz, parolles, & ymages, mais par grandißime elevation d'espirt, & continuel labeur conduict & illustré d'une expresse faveur, & grace de Dieu.', Saint-Gelais, *Advertissement*, sigs. c2r.

role in his creation.[45] Saint-Gelais argued, 'he [God] created them [the stars], as Moses said, not only for marking the times, months, and years, but also with the goal that they serve as signs. Yet there are no signs without signification, and the signification is addressed to people, who alone are able to understand (*entendre*), and not to other animals'.[46] For Saint-Gelais, God's singling out of human beings' obligation to study astrology even extended to the design of their physical bodies. Drawing on Ovid's *Metamorphoses*, he pointed out that people alone among all the animals stood upright so that they could easily view the heavens. God designed other animals so that they were bent over on four legs or even crawled on the earth.[47] Their very body posture meant that God intended human beings to spend their lives in the contemplation of the motions and qualities of celestial bodies and their influences on the earth.[48]

Though focused on the connection between God's creation and the human ability to learn about it through astrology, Saint-Gelais did stress that there were limitations to celestial influences and the human knowledge of these influences. His understanding of these limitations relied on his notions of God's providence and the boundaries of the human mind. He explained that even after their creation, celestial bodies were not entirely independent of God, their creator. God continued to oversee and control the ways that the planets and stars affected the earth through his providence. He offered the biblical example of the city of Nineveh during the lifetime of the prophet, Jonah, to explain this relationship between God's providence and celestial bodies. According to Saint-Gelais, the celestial influences dictated the destruction of Nineveh during this time

---

[45] It is likely that the notion of human beings made in the image and likeness of God as recounted in Genesis 1:26 was behind much of this discussion, but Saint-Gelais did not reference the notion or the biblical passage directly. On this notion and its important in the Renaissance see, Charles Trinkaus, *In Our Image and Likeness: Humanity and Divinity in Italian Humanist Thought*, 2 vols, 2nd ed. (South Bend, IN: University of Notre Dame Press, 1995).

[46] 'Lequel (comme dict Moyse) les crea, non seulement pour compartir les temps, les moys, & les annees, mais außi a fin qu'ilz servissent de signes. Or nul signe n'est sans significantion: & les significantions ne s'adressent que aux hommes, qui seulz les peuuent entendre, et non aux autres animaulx', Saint-Gelais, *Advertissement*, sig. b7v. This was likely a reference to Genesis 1:14. In the Vulgate, Genesis 1:14 reads, 'Dixit autem Deus fiant luminaria in firmamento caeli ut dividant diem ac noctem et sint in signa et tempora et dies et annos'.

[47] Ovid, *Met.* Loeb edition, I.76–88.

[48] Saint-Gelais, *Advertissement*, sig. b8r.

period. The malicious influence of Saturn, Mars, and the other planets
should have destroyed the city. Yet, Jonah's announcement of God's
warning came just in time for Nineveh to save itself through repentance.
The continued existence of the city did not come from any change in
planetary influences but rather through God's relationship to celestial
bodies. As their creator, God directly saved Nineveh from the threat that
Saturn and Mars portended.[49] Though he had designed the celestial bodies
to have predictable effects on the earth during creation, God controlled
how and whether these celestial bodies affected the terrestrial realm in
particular circumstances through his providence.

Saint-Gelais also cautioned people against taking their predictions and
their study of the stars too far. He claimed that contemporary human
knowledge of the world had a limit. His discussion of this boundary again
underscored the connection between God's creation of the world and the
human ability to know it through astrology. According to Saint-Gelais,
even as they used their reason and their upright postures to study the
celestial bodies and make predictions based on their studies, people had to
be very careful to avoid the danger of curiosity. He relied on the examples
of Prometheus and Icarus to explain the outcome of curiosity in the realm
of astrology. For Saint-Gelais, Prometheus' theft of fire from Mount
Olympus and Icarus' misuse of the wax wings that his father had made him
were prototypical examples of those who had pushed their knowledge
beyond acceptable limits. Indulging in curiosity and pride, both
Prometheus and Icarus were punished harshly, and Saint-Gelais warned his
reader to avoid a similar result.[50] The danger of curiosity was so great for

---

[49] Saint-Gelais, *Advertissement*, sigs. b5r–v.

[50] Saint-Gelais, *Advertissement*, sigs. b8v–c1r. For a discussion of these examples
and their use in the various editions of Andrea Alciati's *Emblematum libellus*
(original edition, 1531), see Carlo Ginzburg, 'High and Low: The Theme of
Forbidden Knowledge in the Sixteenth and Seventeenth Centuries', *Past and
Present* 73 (1976): pp. 28–41. On curiosity and its relationship to astrology and
the study of nature more generally, see Hans Blumenberg, *Der Prozeß der
theoretischen Neugierde*, 2nd ed (Frankfurt am Main: Suhrkamp, 1973); Katie
Whitaker, 'The Culture of Curiosity', in *Cultures of Natural History*, ed. N.
Jardine, J. A. Secord, and E. C. Spary (Cambridge: Cambridge University Press,
1996), pp. 75–90; Neil Kenny, *Curiosity in Early Modern Europe: Word
Histories,* Wolfenbüttler Forschungen Band 81 (Weisbaden: Harrassowitz, 1998);
Barbara M. Benedict, *Curiosity: A Cultural History of Early Modern Inquiry*
(Chicago: University of Chicago Press, 2001); Neil Kenny, *The Uses of Curiosity
in Early Modern France and Germany* (Oxford: Oxford University Press, 2004);

him that he explained his reader must take special care to separate true from false knowledge. Attempting to define the boundary between true and false knowledge, Saint-Gelais claimed that there were some mysteries that God reserved for himself and about which it was superstitious and even pernicious for human beings to inquire much as Pico had done. Saint-Gelais made this distinction between true astrology and curiosity in two ways. First, he argued that curiosity appalled the conscience. Second, he connected the illicit or false use of astrology with pagan practices. According to him, pagans explored celestial influences, and subjects such as witchcraft and necromancy that derived their authority only from themselves. True astrologers, on the other hand, knew that God had fashioned the world and that celestial influences ultimately taught people about God who created and continued to control them. Only Christians could practice astrology correctly because they alone knew about God's creation of and continued control of the universe.[51]

Saint-Gelais' defence of predictions of individual human lives made based on the study of celestial bodies drew on many arguments that had appeared in earlier discussion of astrology, as he cited authorities, defended its epistemological status, and separated true astrologers from the untrained who made erroneous predictions. Yet, Saint-Gelais's understanding of God's creation of the world, God's providential control over it, and the human ability to know it provided the ultimate support for astrology in his text. For him, God had designed the planets and stars to affect the terrestrial realm. God had provided people with astrology so that they could understand these influences. God had also fashioned people so that they could and should know this creation. Calling on the studious young woman as well as on the French court in their native language, Saint-Gelais made the study of astrology and the prediction of future events not only permissible but also a duty for all Christians. These Christians could follow this duty as long as they acknowledged God continued to control the celestial bodies and their influence on the terrestrial realm through his providence and as long as they avoided curiosity. Acknowledging God's power over the world and the limits of human knowledge, Christians could properly explore the motions and the positions of the planets and stars and predict future human actions.

---

and R. J. W. Evans and Alexander Marr, eds., *Curiosity and Wonder from the Renaissance to the Enlightenment* (Aldershot: Ashgate, 2006).

[51] Saint-Gelais, *Advertissement*, sigs., b8v–c1r.

John Calvin and his *Advertissement contre l'astrologie judiciaire* (1549)
Also written in the wake of religious reformations, John Calvin's
assessment of astrology in his *Advertissement contre l'astrologie judiciaire*
focused on many of the same themes that appeared in Saint-Gelais' text.
For Calvin as for Saint-Gelais, whether one could make future predictions
based on the knowledge of celestial bodies also depended on how God had
created the world, how God administered the world through his
providence, and how human beings could know about it. Assuming that
God had created the world, Calvin argued that the planets and stars could
only affect the earth if God had designed them to do so and that they could
only have the effects on the earth that God currently allowed them to have.
People could also only possibly know what God had created and what God
had allowed them to know. Much like for Saint-Gelais, Calvin's notions of
creation, providence, and the limits of human knowledge played a
significant role in his assessment of astrology. Yet, he held a different
notion of them and their implications for people and the world they lived in
than Saint-Gelais did. Arguing that God's providence controlled celestial
influences on the terrestrial realm and that human minds could not fully
grasp these effects, Calvin claimed that the study of celestial influences
could not allow people to predict the future course of their lives or world
events.

Rather than relying directly on previous authors of astrological texts as
Saint-Gelais had, Calvin began his assessment of astrology with the
evaluation of its study by the standard of Scripture. He did so to stress the
boundaries God had placed on human knowledge both of God and of the
world because the two were closely connected for him.[52] He framed this

---

[52] On Calvin's notion of the limits of human knowledge especially on the
limitations of the human knowledge of God, see François Wendel, *Calvin: The
Origin and Development of His Thought*, trans. Philip Mairet (New York: Harper
& Row, 1963), pp. 150–165; T. H. L. Parker, *Calvin's Doctrine of the Knowledge
of God*, 2nd ed. (Grand Rapids, MI: Eerdmans Publishing, 1959), pp. 7–13; Edward
A. Dowey, *The Knowledge of God in Calvin's Theology*, 3rd ed. (Grand Rapids,
MI: Eerdmans Publishing, 1994); and Cornelis van der Kooi, *As in a Mirror: John
Calvin and Karl Barth on Knowing God*, trans. Donald Mader (Leiden: Brill,
2005), pp. 1–21. On Calvin's notion of human beings more generally, see Roy W.
Battenhouse, 'The Doctrine of Man in Calvin and in Renaissance Platonism',
Journal of the History of Ideas 9, no. 4 (1948): pp. 447–471; Charles Partee,
Calvin and Classical Philosophy (Leiden: Brill Archive, 1977), pp. 51–65; T. F.
Torrance, *Calvin's Doctrine of Man* (London: Lutterworth Press, 1949); Mary

argument with the biblical verse, 1 Timothy 1:18-19.[53] In the Geneva Bible of 1560, these verses relate Paul's admonition to Timothy to remain an honour and a glory to God forever.[54] Here Paul stated that this would allow Timothy to 'fight a good fight, having faith and a good conscience, which some have put away and as concerning faith, have made a shipwracke (sic)'.[55] His use of the word, 'conscience', in particular suggests the strong link Calvin drew here between astrology and human knowledge of God and of the world. Throughout his writing career in the various editions of his *Institutes* and in discussions of biblical verses that reference the conscience, Calvin often used the word, *'conscientia'*, to refer to a type of innate knowledge, given by God during creation that could help people judge their thoughts and actions much as Saint-Gelais had done in his endorsement of astrology.[56] Calvin addressed the topic of the conscience in his 1540 Commentary on Romans 2:15. He explained:

> *Their conscience also bearing witness and their thoughts accusing one another.* He [Paul] could not have pressed them more strongly than by the testimony of their own conscience, which is equal to a million witnesses. When having done good by the conscience, people sustain and console themselves; those who know themselves to have done evil, they are inwardly tormented and tortured by the conscience. From which those sayings of the pagans come: the good conscience is the largest theater; the evil conscience is truly the worst executioner and more fiercely torments the impious than any furies do. Therefore there is a certain knowledge (*intelligentia*) of the law of nature, which states that this is good and ought to be desired; but that ought to be averted. But observe how learnedly he [Paul] described conscience: he says

---

Potter Engel, *John Calvin's Perspectival Anthropology* (Atlanta: Scholar's Press, 1988); Heiko A. Oberman, 'The Pursuit of Happiness: Calvin between Humanism and Reformation', in *Humanity and Divinity in Renaissance and Reformation: Essays in Honor of Charles Trinkaus*, ed. John W. O'Malley et al. (Leiden: Brill, 1993), pp. 251–283; and Barbara Pitkin, 'The Protestant Zeno: Calvin and the Development of Melanchthon's Anthropology', *The Journal of Religion* 84, no. 3 (2004): pp. 345–378.

[53] Calvin, *Advertissement contre l'astrologie judiciaire,* p. 47.

[54] The translators of this Bible lived intermittently in Geneva alongside Calvin from 1554. The Geneva Bible is therefore the only English translation that was in direct dialogue with Calvin's theology. On the Geneva Bible, see Lloyd E. Berry, 'Introduction to the Facsimile Edition', in *The Geneva Bible: A Facsimile of the 1560 Edition* (Peabody, MA: Hendrickson Publishers, 2007), pp. 1–28.

[55] *The Geneva Bible: A Facsimile of the 1560 Edition*, sig. BBbiv.

[56] See above, pp. 16–17.

that the reasons come to our minds, by which we defend what is done rightly and, on the contrary, that there are those, which accuse and reprove us for our vices. And he directs this accusing and defending to the day of the Lord, not that it will then first happen, for it is now flourishing constantly and carrying out its duty, but since then it will likewise be strong, so that no one should regard it with contempt, as though it were vain and ephemeral. [57]

He later developed this theme further for his 1560 *Institutes*:

All those with just judgment will hold that the human mind (*l'esprit*) has a sense of divinity engraved so deeply that it cannot be erased. And that this persuasion, meaning that there is a God, is naturally rooted in all and that it is fixed in the very bone marrow, is shown by the pride and rebellion of the wicked, who, though they struggle furiously, cannot extricate themselves from the fear of God. Though an ancient by the name of Diagoras and others like him wanted to joke about and scoff at all the religions of the world and though Denis, the tyrant of Sicily, plundering temples, scoffed as if God did not see the drop, the laughter does not escape the throat, for the worm of conscience, keener than burning steel, is gnawing them within.[58]

---

[57] '*Attestante conscientia et cogitationibus*. Non poterat fortius eos premere quam propriae conscientiae testimonio, quae est instar mille testium. Conscientia benefactorum se sustentant ac consolantur homines: male sibi conscii, intra se torquentur ac cruciantur. Unde illae ethnicorum voces: amplissimum theatrum esse bonam conscientiam: malam vero pessimum carnificem, ac saevius quibuslibet furiis impios exahitare. Est igitur naturalis quaedam legis intelligentia, quae hoc bonum *atque expetibile* dictet, illud autem detestandum. Observa autem quam erudite describat conscientiam, quum dicit nobis venire in mentem rationes, quibus quod recte factum est defendamus: rursum quae nos flagitiorum accusent ac redarguant. Rationes autem istas accusandi ac defendendi ad diem Domini confert: non quia sint tunc primum emersurae, quae assidue nunc vigent ac officium suum exercent: sed quia sint tunc quoque valiturae, ne quis ut frivolas et evanidas contemnat.' John Calvin, *Ioannis Calvini Opera Exegetica*, vol. XIII: *Commentarius in Epistolam Pauli ad Romanos*, ed. T. H. L. Parker and D. C. Parker (Geneva: Droz, 1999), p. 47.

[58] Calvin's 1560 *Institutes* I.3.3 reads, 'C'est-cy un point résolu à tous ceux qui iugent iustement, que l'esprit humain a un sentiment de divinité engravé si profond, qu'il ne se peut effacer. Mesmes que ceste persuasion soit naturellement enracinée en tous, assavoir qu'il y a un Dieu, et qu'elle soit attachée comme en la moelle des os, la fierté et rebellion des iniques en testifie, lesquels en combatant furieusement pour se desvelopper de la crainte de Dieu, n'en peuvent venir à bout. Un nommé Diagoras anciennement et quelques semblables ont voulu plaisanter en se moquant de toutes les religions du monde : Denis tyran de Sicile, en pillant les

For both Saint-Gelais and Calvin, the conscience provided people with an objective standard for their behaviour and belief because it came directly from God. Calvin's notion of the conscience differed from Saint-Gelais' in how often Calvin stressed that postlapsarian people tended to ignore or even pollute the judgment of their consciences due to their sinfulness and finitude. Discussing the permissibility of the study of astrology, Calvin pointed out the consequences of polluting the conscience, claiming that this would turn a person away from the pure knowledge of God, leaving him or her to follow lies and errors blindly.

After describing the conscience as a bridle on human knowledge of God, Calvin discussed two other sources of information about him, one of which could also provide people with knowledge of the universe. Though stressing the importance of Scripture for the human knowledge of God, Calvin argued that God had also given people the arts and the sciences so that they could learn about him through the observation and study of the world. According to Calvin, God revealed himself to human beings in the universe. They could perceive this revelation when they understood the world properly through the arts and sciences. Calvin claimed that these were the boundaries God had created for human knowledge and that they alone could lead people to true knowledge of the world. If a person were to transgress these boundaries, he or she would wander aimlessly, ignoring the direct path to knowledge of both the world and of God that God, himself, had established.[59] As he stated when judging those who tried to predict future human actions based on celestial motions:

Even if God had not revealed the purity of his Gospel in our time, nevertheless, seeing that he has revived the human sciences, which are both proper and useful for the guidance of our lives and which, while being used for our benefit, can also serve for his glory, he would still have just cause to punish the ingratitude of those, who, not content with the substantial and well-founded things, crave to flit about in the air, through presumptuous pride.[60]

---

temples s'est moqué comme si Dieu n'y voyoit goutte: mais ces ris ne passent point le gosier, pource qu'il y a tousiours un ver au dedans qui ronge la conscience, voire plus asprement que nul cautère.'

[59] Calvin, *Advertissement contre l'astrologie judiciaire*, pp. 51–53.

[60] 'Quand Dieu ne nous auroit revelé de nostre temp la pureté de son Evangile, veu qu'il a resuscité les science humaines, qui sont propres et utiles à la conduite de notre vie, et, en servant à nostre utilité, peuvent aussi servir à gloire, encore auroit-il juste raison de punir l'ingratitude de ceux qui, ne se contenans point des choses

It is within this context of a good versus a wicked conscience and arriving
at knowledge through a straight path versus blindly wandering in error
over great expanses that Calvin's discussion of astrology took place. Any
licit study of astrology should leave one with a good conscience and lead
one to knowledge of the world and God's revelation in it. The illicit study
of astrology would end in utter ignorance and ultimately, damnation.

   This notion of a limitation on human knowledge of the world and of the
God who revealed himself in it shaped Calvin's judgment of astrology. He
divided it into two different types – natural and judicial. His division
between natural and judicial astrology relied on a distinction between an
art (*ars*) and a science (*scientia*) that was commonly made in late medieval
universities.[61] Sciences or *scientiae* were typically thought to be
knowledge of syllogistically demonstrated conclusions. These conclusions
were understood to be true and certain. The knowledge of a *scientia* was
also thought to be eternal and unchangeable. Therefore, *scientiae* had high
epistemological status. Arts or *artes* did not have the same epistemological
characteristics or status as this eternal and unchangeable knowledge.
Instead, *artes* dealt with practical knowledge of particular things. Focusing
on particulars, the practical knowledge of an *ars* could be changed. Though
both *scientiae* and *artes* contained actual knowledge, *scientiae's* claim to
be unchangeable knowledge lent them more epistemological authority than
the practical knowledge *artes* provided. Calvin explained the *scientia* of
natural astrology thus:

> As well, [natural] astrology serves to ascertain the course of the planets and
> stars, as well for their duration as for their path and position; the duration, I
> say, in order to know (*savoir*) the time each planet and the earth need to finish
> their circuits; the position, in order to judge what is the distance between the
> one and the other, to discern whether the movements are direct or oblique or

---

solides et bein fondées, appetent, par une ambition outrecuidée, de voltiger en
l'air' Calvin, *Advertissement contre l'astrologie judiciaire*, p. 52; Potter, 'A
Warning against Judicial Astrology', p. 163.
[61] See the articles in *Scientia und Ars Im Hoch- und Spätmittelalter*, ed. Ingrid
Craemer-Ruengenberg and Andreas Speer, *Miscellanea Mediaevalia*, Vol, 22
(Berlin: Walter de Gruyter, 1994); see also Charles H. Lohr, 'Aristotelian
'Scientia', the 'Artes', and 'English Philosophy in the 14th Century', in *Erkenntnis
und Wissenschaft: Probleme der Epistemologie in der Philosophie des Mittelalters
/ Knowledge and Science: Problems of Epistemology in Medieval Philosophy*, ed.
Matthais Lutz-Bachmann, Alexander Fidora, and Pia Antolic (Berlin: Akademie
Verlag, 2004), pp. 265–273.

quasi-contrary; to be able to show why the sun is further from us in the winter than in the summer and why it stays with us longer in the summer than in the winter; to be able to use a compass to determine what sign of the zodiac a planet or star occupies each month, and what intersections it has with other planets; to know why the moon waxes or wanes as it recedes from and approaches the sun; to understand how eclipses happen and to be able to mark their positions' degree and minute with the compass.[62]

Though those who actually studied the celestial bodies seriously had more knowledge of this *scientia*, Calvin also claimed that those without such education had rudimentary knowledge of it. After all, all people could see that the celestial bodies moved, that day alternated with night, and that the seasons changed, even if they did not understand how such things took place in great detail.[63]

Having argued that even those who could just view the heavens had some knowledge of this *scientia* and after explaining what that *scientia* encompassed, Calvin described the *ars* of natural astrology thus: 'With this foundation [of the *scientia*] laid, the effects, which we see on the earth, follow. It is through [natural] astrology that we know they come from above...'[64] For him, the planets did have an effect on the lower, earthly realm that the study of natural astrology helped people understand. Yet, he restricted these effects to the properties and motion that God had set into the planets and stars at creation. According to him, these natural properties could affect the earth in two different ways. First, he explained that the

---

[62] 'Par ainsi, l'astrologie sert à determiner le cours des planettes et estoilles, tant pour le temps que pour l'ordre et situation; le temps, dy-je, pour savoir quel terme il faut à chascune planette et au firmament pour accomplir leur circuit; la situation, pour juger combien il y a de distance de l'une à l'autre, discerner les mouvemens droitz ou obliques, ou quasi contraires; de là, savoir monstrer pourquoy le soleil plustost est plus loing de nous en hyver qu'en esté, pourquoy il fait plus longue demeure sur nous en esté qu'en hyver; de savoir compasser à l'endroit de quel signe du zodiaque il est chascun moys, quelles rencontres il a avec les autres planettes, pourqouy la lune est pleine ou vuyde selon qu'elle se recule du soleil ou en approache, comment se font les eclypses, voire jusques à compasser les degrez et minutes'. Calvin, *Advertissement contre l'astrologie judiciaire*, pp. 54–55. Potter, 'A Warning against Judicial Astrology', p. 165.

[63] Calvin, *Advertissement contre l'astrologie judiciaire*, p. 54.

[64] 'Ce fondement mis, s'ensuyvent les effectz que nous voyons icy bas, lesquelz par l'astrologie on cognoist provenir d'en haut...', Calvin, *Advertissement contre l'astrologie judiciaire*, p. 55. Potter, 'A Warning against Judicial Astrology', p. 165.

motion of the heavens, acting as God's instruments, helped cause
meteorological phenomena such as wind, rain, hail, snow, and earthquakes.
With a proper knowledge of the *scientia* of planetary motion, a person
could predict some of these phenomena, although they always remained
subject to God's direct intervention.[65] In addition to meteorological
phenomena, Calvin also argued that the planets could have an impact on
the dispositions of human bodies because the planets could influence the
balance of the qualities of hot or cold, dry or wet within them. In this way,
natural astrology could be useful to medical practitioners because
knowledge of a patient's qualities would allow them to decide when to
practice bloodletting or administer medications.[66] Based on the *scientia* of
the heavenly bodies, farmers and medical practitioners could learn to
predict when to plant their crops and how best to treat their patients
through the *ars* of natural astrology because God had created celestial
bodies to effect meteorological phenomena and human bodies. Though
natural astrology was a valid application of this *scientia*, Calvin argued
strenuously against the type of astrology he called, 'judicial'. He claimed
that judicial astrology consisted of two things: trying to predict the fortunes
of an individual and all that he or she would do or experience, and trying to
predict the course of events for the entire world.[67] He attacked judicial
astrology from two perspectives. The first perspective involved an attack
on its actual ability to predict anything about either individuals or world
events. This critique echoed many traditional, negative assessments of the
study of celestial influences and its faulty epistemological status, especially
those of Augustine of Hippo, whose examples Calvin did adapt for his
text.[68] Calvin's ultimate critique of judicial astrology, however, had little to
do with its inability to predict the future course of human lives or world
events. It also had little to do with its epistemological presumption.
Toward the end of the text, he returned to the theme with which he had
begun his argument – the limitations on the human knowledge of the
world. According to Calvin, Christians had to avoid judicial astrology
because it was a transgression of the boundaries that God had established
for their knowledge. God did not want people to worry about the future
when they viewed the heavens because he actively continued to direct what

---

[65] Calvin, *Advertissement contre l'astrologie judiciaire*, pp. 55–56.
[66] Calvin, *Advertissement contre l'astrologie judiciaire*, pp. 56–57.
[67] Calvin, *Advertissement contre l'astrologie judiciaire*, p. 57.
[68] Calvin, *Advertissement contre l'astrologie judiciaire*, pp. 57–69.

took place in the universe through his providence.[69] Calvin drew proof of this directly from biblical verses. Most of these biblical proof texts came from the prophets Isaiah and Jeremiah. The vast majority of them attacked the teachings of Egyptians and Chaldeans, assumed by most sixteenth-century writers to be the originators of astrology.[70] Jeremiah 10:2 is the summation of Calvin's argument. In the Geneva Bible, the verse reads, 'Thus saith the Lord: Learn not the way of the heathen, and be not afraid for the signs of heaven, though the heathen be afraid of such'.[71] God here forbade the Israelites to practice 'the ways of the heathen.' For Calvin, in doing this, God had also forbidden the practice of judicial astrology. People were not to be afraid of what they saw in the heavens because people should and could not know their futures. Instead, they had to trust in God's promise that he would continue to direct their lives and the occurrences on the earth. As Calvin stated,

> I know well the subterfuge that some bring forth that one need not fear the stars as if they held dominion over us even though they do have some subaltern superiority [for affecting individual human lives] under the control and guidance of God. But there is no doubt that the prophet [Jeremiah] wants to lead us to the providence of God.[72]

---

[69] For a discussion of Calvin's notion of providence, see Josef Bohetec, 'Calvins Vorsehungslehre', in *Calvinstudien Festschrift zum 400. Geburtstag Johann Calvins* (Leipzig: Rudolf Haupt, 1909), pp. 339–441; W. J. Torrance Kirby, 'Stoic *and* Epicurean? Calvin's Dialectical Account of Providence in the *Institutes*', *International Journal of Systematic Theology* 5, no. 3 (2003): pp. 309–322; and Pieter C. Potgieter, 'Providence in Calvin: Calvin's View of God's Use of Means (*media*) in His Acts of Providence', in *Calvinus Evangelii Propugnator: Calvin, Champion of the Gospel*, ed. David F. Wright, et al. (Grand Rapids, MI: CRC Product Services, 2006), pp. 175–190. For a discussion of Calvin's understanding of providence in relationship to his understanding of creation, see Richard Stauffer, *Dieu, la création et la Providence dans la prédication de Calvin* (Berne: Peter Lang, 1978).

[70] On this assumption, see Sylviane Bokdam, 'Les mythes de l'origine de l'astrologie,' in Aulotte, *Divination et controverse religieuse en France au XVIᵉ siècle*, pp. 57–72.

[71] *The Geneva Bible: A Facsimile of the 1560 Edition*, sig. Gggiiiiv.

[72] 'Je say bien le subterfuge qu'amenent aucuns, qu'il ne faut pas craindre les astres comme s'ilz avoyent domination sur nous, et toutefois qu'ilz ne laissent point d'y avoir quelque superiorité subalterne, souz la main et conduite à Dieu. Mais il n'y a nulle doute que le prophete ne vous vueille r'amener à Providence de

Focusing on God's providential control over the world kept people from attempting to fly through the air in a vain quest for knowledge of God and the world that they could never hope to attain. More importantly, avoiding such vain and dangerous curiosity and staying within the limits God had set for human inquiry, people could actually gain the knowledge of God that he had chosen to reveal in the world.[73] This emphasis on a licit way of knowing the world also sheds light on Calvin's endorsement of natural astrology. Similar to his critique of judicial astrology, Calvin did not support natural astrology just because he believed it could yield correct information. Instead, he advocated its study because God himself had sanctioned its *scientia* and *ars* as a way to discover his continued intervention into the universe through his providence. To understand this, it is important here to recall what Calvin said about the sources of human knowledge of God at the beginning of his *Advertissement*. He claimed that there were three noetic paths to God – a good conscience, heavenly doctrine, and the liberal arts and sciences. In the *Institutes* and in his *Commentary on Romans* (1540), Calvin certainly expressed much distrust of both the human conscience and the human intellect because, he claimed, postlapsarian people were too sinful to use them both properly. The Word and God's guidance of the elect alone could provide people with sure knowledge of God. Nevertheless, in these works, Calvin repeatedly stated that God had revealed aspects of himself in the world whether or not sinful humanity was capable of seeing them there. He also held out the possibility that with the training of Scripture and the grace of God, postlapsarian people could once again use the liberal arts and sciences as a way to see God's revelation in the world, even if they could only glimpse glowing embers and small sparks of it. These glimpses should then lead people to praise their creator.[74] Assuming a person had God's aid, natural astrology was one sanctioned liberal art that could help a person learn about God's creation and his status as its creator. In particular, natural astrology could assure people that God would and could keep his promise to save humanity because it provided the knowledge that he continued to intervene actively

---

Dieu...' Calvin, *Advertissement contre l'astrologie judiciaire,* pp. 75–76. Potter, 'A Warning against Judicial Astrology', pp. 177–178.

[73] Calvin, *Advertissement contre l'astrologie judiciaire,* pp. 73–80.

[74] Calvin developed this theme at length in the 1545 *Institutes*. See his *Institution de la religion chrestienne: compose en Latin par Jehan Calvin, & translate en Francoys par luymesme: en laquelle est comprise une somme de toute la Chrestienté* (Geneva: Jean Girard, 1545), sigs. a5r–b1r.

into the world he had created through his providence. Building on Aristotelian notions of the heavens' order, stability, and harmonious circular motion, Calvin argued that these obvious signs of perfection revealed some of God's attributes.[75] Celestial bodies' perfect motion and complete stability testified to God's power, wisdom, and goodness because he alone had the power to uphold these bodies in their perfection.[76] Drawing on Psalm 19, Calvin stated, 'In short since the heavens show us the praise of God and the earth provides us with a mirror of his virtue and infinite wisdom, let us, to do much for our profit, never wander about in our imagination, which carries us away from him'. [77] Lest people fear the impact that these heavenly bodies could have on meteorological phenomena, Calvin again pointed out that God was ultimately in control of these phenomena and only used the planets as his instruments.

> Now I confess well, following what I said before that terrestrial bodies have some correspondence with the heavens, one can note well certain causes in the stars for things that happen here below. Just as the influence of the heavens often causes tempests, whirlwinds, changes in the weather and continual rain, so it also consequently brings about sterility and pestilences. In as much as one sees an order and a sort of liaison between the higher and the lower, I do not forbid anyone to search the heavenly creatures for the origin of the accidents that we see in the world; I understand the 'origin,' not as the principle but as an inferior means to the will of God...[78]

---

[75] Aristotle, *On the Heavens*, Revised Oxford Translation edition, 1.2.269a14–1.3.270b31. See also Edward Grant, *A History of Natural Philosophy: From the Ancient World to the Nineteenth Century* (Cambridge: Cambridge University Press, 2007), pp. 37–42.

[76] Calvin, *Advertissement contre l'astrologie judiciaire*, pp. 80–82.

[77] 'Brief, puis que les cieux nous doyent raconter la louange de Dieu, et le firmament nous doit estre un miroir de sa vertu et sagesse infinie, apprenons, pour en bien faire nostre profit, de ne point extravaguer en des imaginations qui nous eslongnent de luy.' Calvin, *Advertissement contre l'astrologie judiciaire*, p. 82. Potter, 'A Warning against Judicial Astrology', p. 180.

[78] 'Or je confesse bien, suyvant ce que j'ay cy dessus traité, qu'entant que les corps terrestres ont convenance avec le ciel, on peut bien noter quelque cause aux astres des choses qui avicnnent icy bas. Car tout ainsi que l'influence du ciel cause souvent les tempestes tourbillons et temps divers, item les pluyes continuelles, ainsi, par consequent, elle amene bien la sterilité et pestilences. Entant donc qu'on verra un ordre et comme une laison du haut avec le bas, je ne contredy pas qu'on ne cerche aux creatures celestes l'origine des accidens que on void au monde, j'entens l'origine non pas principale, ains comme moyen inférieur à la volonté de

For Calvin, God alone ultimately caused droughts to punish the wicked and chastise the faithful. God alone sent abundant rain to water fields and caused crops to grow so that people did not starve. God's ability both to correct and to provide for people revealed his care for them. Whether one was a student of natural astrology or one of the simple and unlearned people Calvin claimed to be writing for, each and every one had the capacity to observe the planetary bodies, the stars, and their motions, to find in them testimony that God had the power and the will to save them from eternal death, and to praise God for this testimony and care. At the end of his text, Calvin laid out some remedies to keep students of natural astrology from sliding into judicial astrology without realizing it. This discussion again reveals that the themes of the limitations of human knowledge and God's continued control of the universe through his providence even after creation were basic to Calvin's assessment of astrology. Calvin proposed that these students of natural astrology should always practice their art with the fear of God in mind. Profitable study should educate people in such fear. It should also allow them to retain a good conscience. As Calvin asked,

> What remedy, then, should we prescribe to cure such evils? It is that sobriety that Saint Paul recommends to us because it acts as a bridle to hold us in the pure obedience of God. In order to practice it, let each person be advised to guard well that inestimable treasure of the Gospel in good conscience because it is certain that the fear of God will be for us a good rampart, defending us against all errors.[79]

Natural astrology practiced in the fear of God would lead practitioners to use the art to discover God's creation of and continued control over the universe through his providence and to teach this knowledge to others so that they, too, could be edified. Avoiding the vain curiosities of judicial astrologers and teaching natural astrology to others, these scholars would

---

Dieu...', Calvin, *Advertissement contre l'astrologie judiciaire*, pp. 69–70. Potter, 'A Warning against Judicial Astrology', p. 174.

[79] 'Quel remede donc pour obvier à telz inconveniens? C'est que la sobriété que sainct Paul nous recommende nous soit comme une bride pour nous tenir en la pure obeissance de Dieu, et, pour ce faire, que chascun advise bien de garder ce thresor inestimable de l'Evangile en bonne conscience. Car il est certain que la criante de Dieu, sera un bon rempar pour nous munir contre tout erreurs.' Calvin, *Advertissement contre l'astrologie judiciaire*, pp. 69–70. Potter, 'A Warning against Judicial Astrology', pp. 188–189.

give people confidence in God's promises, and it would provide a bridle to keep them obedient to God.[80] Anxious to retain God's control over the world as well as to warn his readers away from curiosity, Calvin ultimately condemned judicial astrology in his *Advertissement contre l'astrologie judiciare* because he claimed it threatened to lead people into sin and impiety as they exceeded the limits God had set for their knowledge of him and of the world.

Mellin de Saint-Gelais and John Calvin on Astrology
This investigation of Calvin's *Advertissement contre l'astrologie judiciaire* alongside Mellin de Saint-Gelais' *Advertissement sur les iugemens d'astrologie à une studieuse damoyselle* places Calvin's condemnation of Charles de Dançay's use of astrology from 1539–1541 in a new light. Calvin did not chastise Dançay because he was a theologian slavishly following the age-old tradition of Christian theologians. Such a chastisement would not be compatible with Calvin's endorsement of natural astrology. Instead, Dançay had tried to use the methods of judicial astrology to predict the future actions of individual women. By turning to the heavens to find out whether these women would commit adultery, Dançay sought knowledge that Calvin argued celestial bodies could not reveal. More importantly, he also attempted to do something that Calvin argued God had forbidden – to know the future paths of individuals that only God alone had determined. Despite his previous practices and opinions, Dançay also endorsed a separation between studying celestial bodies and making such predictions based on these studies in his 1576 letter to Tycho Brahe. Praising him for his infrequent use of this kind of astrology, Dançay complemented Brahe's work and expressed the hope that Brahe would be willing to share his theory of planetary motion with him.[81] In so far as Brahe's theory of planetary motion could, according to Dançay, reveal aspects of God's continued interaction with the universe through his providence and the limited character of the human knowledge of the world, one could perhaps imagine Saint-Gelais and even Calvin eager to receive this information as well. Their treatises show a deep concern with how God continued to relate to the created universe through his providence and the human beings whom they both argued God intended to know the world. They both wrote frequently of God's direct intervention into the world to alter or even change something that he had

---

[80] Calvin, *Advertissement contre l'astrologie judiciaire*, pp. 94–102.
[81] Dançay, 'Carolus Danzaeus ad Tychonem 17 Nov. 1576', p. 41.

shaped during creation. They also focused on how God had fashioned human beings so that they could learn about the universe and the God who continued to be directly involved with it. The Christian obligations to observe and study the world to learn about God and to praise him as its creator were essential to both Calvin and Saint-Gelais' judgments of astrology.

Though analysing similar topics to assess the merits of astrology, Saint-Gelais and Calvin's different conceptions of how God had fashioned the world, how he continued to relate to it, and how human beings could know it, led them to evaluate it differently. Whereas Saint-Gelais held that God had designed celestial bodies so that they had predictable effects on the earth and individual human lives and so that human beings could predict these effects, Calvin argued that this was not how God had created the world. For him, heavenly bodies did not determine the course of individual human lives nor could limited and often blind human minds grasp all aspects of celestial influences. Their judgments of astrology incorporated questions surrounding creation, providence, and human knowledge. Their focus on these themes in their assessments of astrology suggests that the very understandings of creation, providence, and the human knowledge of the world might have been in question for other sixteenth-century authors as they, too, experienced the pressures of religious reformations on traditional Christian doctrines and the impact that these pressures had on notions of the universe.

The Contextual Rationality of Galileo's Astrology

Scott Hendrix

Abstract: In order to understand the relationship between Galileo's work and his astrological interests, it is first necessary to understand the socio-historical contexts of the rationality of his astrological beliefs. Many modern people view a belief in predictive astrology as inherently irrational and in opposition to a scientific worldview, as evidenced by scholars such as Karl Popper and scientists such as Carl Sagan. However, drawing on the work of the anthropologist Steven Lukes, the philosopher Richard Swinburne, and the sociologist Pierre Bourdieu, this article argues that the intellectual worldview that generations of European scholars created during the Middle Ages, Renaissance, and early modern periods created a context in which an acceptance of astrological principles was part of the rational fabric of the seventeenth century. Therefore, any evaluation of Galileo's work, including his astrological interests, should use that context as a starting point in order to avoid modernist biases.

Since the time of Lynn Thorndike and Dame Frances Yates serious scholarly studies of the place of astrology and astrologers in history have become both more common and increasingly rigorous in their approach. Laura Ackerman Smoller's *History, Prophecy, and the Stars* hooked me on the study of astrology's history, while Paola Zambelli's volume on *The Speculum Astronomiae and its Enigma* should be the starting point for anyone interested in medieval astrology and Anthony Grafton's highly readable *Cardano's Cosmos* is destined to become a classic. Nevertheless, one still encounters unsupportable and ahistorical attitudes toward the history of astrology. Even such fine scholars as Louis Dupré all too often approach premodern astrological beliefs as barriers to the development of modern – and by this term Dupré and others mean rational – modes of thinking.[1] It is not hard to uncover the rationale for this approach: the philosophers of science Boris Castel and Sergio Sismundo have said, 'The modern world, and perhaps what it means to be modern, is thoroughly

---

[1] Louis Dupré, *Passage to Modernity: An Essay in the Hermeneutics of Nature and Culture* (New Haven: Yale University Press, 1995), pp. 56–57.

Scott Hendrix, 'The Contextual Rationality of Galileo's Astrology', *Culture and Cosmos,* Vol. 18, no. 2, Spring/Summer 2014, pp. 71-103.
www.CultureAndCosmos.org

entwined with science',[2] and in the modern world the terms 'rational' and 'scientific' are used as near synonyms.[3] Therefore, that which is not 'scientific' cannot be 'rational', and modern scientists like Carl Sagan and philosophers of science such as Karl Popper are quick to point to astrology as the epitome of a pseudo-scientific belief.[4]

Unfortunately, these views of astrology as 'unscientific' or 'pseudoscientific', with the corresponding connotations of 'irrational', frequently bleed over into historical studies. Most historians have come to understand that premodern Europeans typically did not question the precepts of natural astrology, which focuses on events in nature such as weather, even if there were those during the Middle Ages and beyond who found judicial astrology – the application of astrological principles to make predictions about the future or find out information about a person's health or when the most propitious time to undertake an activity might be – questionable or downright objectionable. However, other scholars lack even this basic understanding of the differences in the way people viewed astrology. For example, Jean-Pierre Torrell muddles natural and judicial astrology under the blanket condemnation of 'astrological superstitions', applied in an entirely ahistorical way in his study of *St. Thomas Aquinas: The Person and His Work*.[5] In stark contrast, David Pingree has described the work of Babylonian astrologers as 'science'.[6] Why the confusion? Not to be simplistic, but these problems have arisen because we need to re-evaluate what it means to be rational. Additionally, we must avoid

---

[2] Boris Castel and Sergio Sismondo, *The Art of Science* (Toronto: UTP Higher Education, 2003), p. 9.

[3] For an in-depth analysis of this issue, see Scott E. Hendrix, 'Natural Philosophy or Science in Premodern Epistemic Regimes? The Case of the Astrology of Albert the Great and Galileo Galilei.' *Teorie vědy / Theory of Science: Journal for Interdisciplinary Studies of Science*, XXXII.2 (2011): pp. 111–132.

[4] Karl Popper, *Objective Knowledge: an Evolutionary Approach* (Oxford: Oxford University Press, 1975), p. 176; Carl Sagan, 'The Harmony of the Worlds,' *Cosmos*, PBS, 1980.

[5] Jean–Pierre Torrell, O.P., *St. Thomas Aquinas: The Person and His Work*, trans. Robert Royal (Washington, D.C.: The Catholic University of America Press, 2005, 2nd printing), p. 215. In contrast, see Scott E. Hendrix, 'Contemplating the Stars and Comprehending Humanitas: Thomas Aquinas on Astrology and the Essence of what it is to be Human,' *History and Culture: Essays on the European Past*, ed. Nicholas C. J. Pappas (Athens: ATINER, 2012): pp. 11–22.

[6] David Pingree, 'Astrology', in *Religion, Learning, and Science in the 'Abbasid Period*, ed. M. J. L. Young, J. D. Latham, and R. B. Serjeant (Cambridge: Cambridge University Press, 1990), pp. 290–99.

piecemeal analyses of premodern intellectuals that seek to separate what is
'scientific' and thus 'rational' from that which is neither. It is just this sort
of confusion that has caused otherwise fine researchers such as Eugenio
Garin to perform feats of intellectual gymnastics in order to explain away
what might appear to be 'superstition' when encountered in the work of
Galileo Galilei (1564–1642).[7] However, such efforts are as unjustified as
they are unnecessary, for Galileo worked as a judicial astrologer early in
his life and supported astrological ideas throughout his life. Why should he
not have done so? His interest in astrology was rational given the
intellectual and sociohistorical contexts within which he worked, and his
astrological interests may well have reinforced, or at least did not impede,
the habits of mind that promoted a burgeoning scientific approach in his
work. Bernadette Brady has eloquently argued this latter point.[8] An
analysis of contextual elements of the rationality of Galileo's astrological
beliefs will enhance an understanding of the positive contributions of his
astrological beliefs, while acting as a further corrective to other distortions
of his career caused by the 'Galileo legend', which refers to the complex
web of myths that have grown up around Galileo. Although many scholars
have noted the distorting effects of these myths, astrology has not yet been
a point of interest in the discussion.[9]

Let us begin with those who do not like to think of Galileo as an
astrologer. Some prefer to sidestep the issue, accepting that Galileo did
astrological work early on but then turned against this 'superstition', which
is at times portrayed as a foundational step in the development of science.
A good example of this approach can be found in the work of David
Wooten, who argues that Galileo's rejection of the principles of causation
underpinning astrology's claims came about after his work with floating
bodies; this work weakened acceptance of influence at a distance in the
eyes of many, and therefore could have reduced support for astrology's
basic principles. That is Wooten's position, and for him Galileo's

---

[7] Eugenio Garin, *Astrology in the Renaissance: The Zodiac of Life*, trans. Carolyn
Jackson and June Allen (New York: Routledge, 1983), p. 10.
[8] Bernadette Brady, 'Galileo's Astrological Philosophy', in *From Masha'Allah to
Kepler: Theory and Practice in Medieval and Renaissance Astrology*, ed. Charles
Burnett and Dorian Gieseler Greenbaum (Ceredigion: Sophia Centre Press, 2015),
pp. 77–100.
[9] Phillip Kitcher, *The Advancement of Science: Science without Legend,
Objectivity without Illusions* (Oxford: Oxford University Press, 1993), pp. 230–
233; T. S. Lessl, 'The Galileo Legend as Scientific Folklore,' *Quarterly Journal of
Speech* 85, no. 1 (1999): pp. 146–68.

denunciation of astrology was a major step in the development of his scientific modes of thought.[10] Wooten finds this denunciation in a cast-off comment Galileo inserted into the mouth of Salviati in his famed *Dialogue Concerning the Two Chief Systems of the World*, published in 1632. The character of Salviati frequently expresses Galileo's own views, and he notes that the 'prophecies of astrologers... are so clearly seen in horoscopes... after their fulfillment'.[11] In other words, judicial astrologers are frauds who 'read back' and demonstrate that horoscopes they had previously cast did in fact predict events as they came to pass.

Wooten is not the only scholar to find the denunciation placed in the mouth of Salviati to be a statement regarding Galileo's own views of judicial astrology. As Germana Ernst notes, Galileo – speaking through Salviati – joins the 'predictions of natal astrology to the prophecies of Joachim of Flore, and to the responses of the oracles of the Gentiles' in his dismissal of prophecies that 'only make sense after the prophesised events'.[12] However, denunciations of frauds were a commonplace among judicial astrologers and those who wrote about the discipline. Michael D. Bailey argues that this habit has confused historians on numerous occasions.[13] Additionally, we should be careful how we interpret statements that seem to be suggesting a rejection of astrological beliefs. For example, when Tommaso Campanella (1568–1639) chastised Galileo in a letter dated 8 March 1614 for his 'disbelief' of astrological medicine when the latter refused to accept an astrological consultation for a health condition, that is an expression of frustration on the part of Campanella indicating his interpretation of the reason for Galileo's stubbornness on this matter.[14] We cannot assume that Campanella was right about Galileo's

---

[10] David Wooten, 'Galileo: Reflections on Failure', in *Causation and Modern Philosophy*, ed. Keith Allen and Tom Stoneham (New York: Routledge, 2011): pp.13–30, pp. 21–23.
[11] Galileo Galilei, *Dialogue Concerning the Two Chief Systems of the World*, trans. and ed. Stillman Drake (Berkeley: The University of California Press, 1953), pp. 109–110.
[12] Germana Ernst, 'Astrology and Prophecy in Campanella and Galileo', *Galileo's Astrology*, special issue of *Culture and Cosmos* 7, no. 1 (2003): pp. 21–36, p. 33.
[13] Michael D. Bailey, *Fearful Spirits, Reasoned Follies: The Boundaries of Superstition in Late Medieval Europe* (Ithaca: Cornell University Press, 2013), p. 131.
[14] Germana Ernst discusses this interchange in her essay, 'Gli Astri e la Vita del'Uomo: Gli Opuscoli Astrologici di Tommaso Campanella', *Nella luce degli astri. L'astrologia nella cultura del Rinascimento,* ed. Ornella Pompeo Faracovi

motives. Furthermore, Michael A. Ryan is currently working on a project investigating instances of magical and astrological fraud in the Mediterranean region during the Middle Ages. Fraud appears to have been common among those practicing such disciplines, a fact that was readily known to professional astrologers, who took understandable umbrage at their unscrupulous peers.[15] Read in that context, Galileo's comment is a denunciation of frauds, rather than a rejection of astrology as a discipline, which had been Favaro's position when he first wrote of Galileo's astrology in 1881.[16] Therefore, Wooten seems to be reading too much into Galileo's obscure comment.

It is unclear why Wooten takes his position, but Joseph Agassi's refusal to acknowledge Galileo's work as an astrologer is easier to explain. Agassi is a prominent philosopher of science whose work is a feature of many philosophy of science classes, and his book provides a good example of those who simply ignore any association between Galileo and astrology. This approach is understandable, given that he refers to Lynn Thorndike's magisterial *History of Magic and Experimental Science* as 'useless and infuriating' while contending that 'black superstitions' such as astrology retarded the development of science. Agassi even goes so far as to vigorously oppose efforts to analyse or record such 'superstitious' beliefs of the past.[17] Therefore, Agassi has much to say of what he views as Galileo's rational approach to the world and nothing at all to say about his work as an astrologer. Ignoring Galileo's work as an astrologer is, in fact, the norm. As just one example, in 476 pages *The Cambridge Companion to Galileo* does not mention astrology once.[18]

Wooten and Agassi serve as examples of the scholarly position that denies or ignores Galileo's involvement with astrology. What, then, is the evidence for Galileo as an astrologer, rather than the 'constant adversary of

---

(Sarzana: Agorà, 2004): pp. 5–57, pp. 7–8; Germana Ernst, 'Aspetti dell'astrologia e della profezia in Galileo e Campanella'. in *Novità celesti e crisi del sapere*, ed. P. Galluzzi (Florence: Florence: Barbera, 1984): pp. 255–266.

[15] Michael A. Ryan, 'A Note on Magical Deception', *Magic, Ritual, and Witchcraft* 7, no. 1 (2012): pp. 52–57.

[16] Antonio Favaro, 'Galileo Astrologo', trans. Julianne Evans, in *Galileo's Astrology*, special issue of *Culture and Cosmos* 7 no. 1 (2003), pp. 9–19, p. 14.

[17] Joseph Agassi, *Science and its History: A Reassessment of the Historiography of Science* (New York: Springer, 2008, 2nd edition), pp. 134–135.

[18] Peter Machamer, ed., *The Cambridge Companion to Galileo* (Cambridge: Cambridge University Press, 1998).

divinatory astrology' Eugenio Garin portrays him to be?[19] Such evidence began to emerge as long ago as 1881, when the editor of Galileo's collected works, Antonio Favaro, published an article entitled 'Galileo Astrologo,' which was translated and included in a special volume on *Galileo's Astrology* released by *Culture and Cosmos* in 2003.[20] Although Favaro's conclusions were tentative and he suggested that perhaps Galileo had lost interest in astrology as he aged, the evidence presented leaves the reader with no doubt that Galileo cast horoscopes for many years. This evidence ranges from the natal chart (a horoscope predicting an individual's future based on his or her time of birth that Galileo cast for the Grand Duke Cosimo II of Florence, to epistolary discussions of horoscopes in which Galileo engaged with important but distant figures such as the Cardinal Alessandro d'Este as well as close friends such as Giovanni Francesco Sagredo.[21] One would think that such evidence would be compelling, but even historians who acknowledge the existence of these horoscopes have not given up efforts to explain them away. Richard Tarnas, for example, argues that Galileo's astrological work was intended to garner patronage from powerful people such as the Dukes of Florence.[22] Wade Rowland quickly passes over it as an activity in which Galileo 'dabbled'.[23] These efforts to minimize or ignore Galileo's interest in astrology have been facilitated by the fact that most of the manuscript charts he drew up remain unpublished. As Germana Ernst points out, the roughly fifty folios of astrological charts Galileo produced have been reduced to two printed pages, which reproduce the titles and the most discursive elements, but leave out both the charts themselves as well as Galileo's copious notes on these charts.[24]

However, if we take the time to examine the horoscopes Galileo cast there seem to be good reasons to accept that he saw judicial astrology as a

---

[19] Garin, *Astrology in the Renaissance*, p. 10.

[20] Favaro, 'Galileo Astrologo', pp. 9–19.

[21] Favaro, 'Galileo Astrologo', pp. 11–13.

[22] Richard Tarnas, *The Passion of the Western Mind* (New York: Random House, 1993), p. 295.

[23] Wade Rowland, *Galileo's Mistake: A New Look at the Epic Confrontation Between Galileo and the Church* (New York: Arcade Publishing, 2003), p. 295.

[24] Ernst, 'Astrology and Prophecy in Campanella and Galileo', p. 29. The charts in Galileo's hand may be found in Florence at the Biblioteca Nazionale Centrale, mss. Galil. 81. Cf. Galilei, *Opere*, XIX, pp. 205, 206, 218–220. The published titles of Galileo's charts and their discursive elements can be found under the title *Astrologica Nonnulla*. See Galileo Galilei, *Opere*, XIX, pp. 205, 206, 218–220.

useful and effective tool. As Darrell Rutkin points out, the horoscopes Galileo cast include not only those that could have been intended to garner the support of a patron – such as the aforementioned one for Cosimo II – but also those Galileo did for himself, his daughters, and twenty as yet to be identified people.[25] Additionally, Rutkin has analysed the natal charts Galileo completed for his daughters and his friend Giovanni Francesco Sagredo, noting the attention to detail evident in these charts. The purpose of such charts could not possibly have been to earn a fee or garner patronage, leaving us to see the care with which Galileo approached his work as an indication of the respect for the discipline in which he was engaged. The chart Galileo did for himself provides a good example of the careful approach he took to the astrological casting, containing as it does numerous corrections, indicating that he carefully worked and reworked this chart to ensure its accuracy.[26] True, it is possible that he made these charts simply to practice casting horoscopes, but then why would he take such extreme care? It is not as if his daughter would question his mathematics.

Clearly, the existence of this sizeable number of carefully constructed horoscopes in Galileo's own hand demonstrates not only that he was quite familiar with the workings of judicial astrology, but also that when he did construct a chart he demonstrated both a great deal of care as well as ability. It is true that these horoscopes all seem to date to the late 1580s and 1590s while he was a professor of mathematics at Padua, but there is no evidence that he changed his mind about the discipline after his appointment to the position of 'Chief Mathematician of the University of Pisa and Philosopher and Mathematician to the Grand Duke' of Tuscany in 1610, and indeed as we will see later he continued to insert astrological references into his work down to 1632. Yes, there are statements such as that in his *Dialogue on the Two Chief System of the World* that join 'the predictions of natal astrology to the prophecies of Joachim of Fiore, and to the responses of the oracles of the Gentiles'.[27] This, however, is evidence of an impatient and rejectionist attitude toward prophecy, as well as unlearned or poorly worked through astrology, that is common to professional astrologers of the medieval and early modern periods. It is

---

[25] H. Darrel Rutkin, 'Galileo Astrologer: Astrology and Mathematical Practice in the Late Sixteenth and Early Seventeenth Centuries', *Galilaena* II (2005): pp. 107–143.

[26] This chart, along with Deborah Houlding's commentary on it, is available at: http://www.skyscript.co.uk/galchart.html [accessed 23 September 2013].

[27] Ernst, 'Astrology and Prophecy in Campanella and Galileo', p. 33.

also true that he no longer produced astrological predictions, but the simplest explanation for that lapse is that other concerns simply kept him too busy for such pursuits, especially since he became increasingly anxious to solidify his status as a philosopher rather than simply a professor of mathematics.[28]

In the absence of evidence that Galileo came to reject the ideas underpinning astrology, we should ask why scholars have attempted to distance him from the discipline. To understand that, we must begin by understanding the preconceived position promoted by a modernist worldview.[29] If what it is to be 'modern' is to be 'scientific,' as Castel and Sismundo have stated,[30] and 'rational' and 'scientific'[31] are all but equated within this worldview, then we begin to glimpse the historical blinkers that have been hampering our understanding of Galileo. Immediately after advancing his famous definition of science, that it must make risky predictions that are falsifiable,[32] Karl Popper then provides astrology as an example of a pseudoscience, the paradigmatic example of that which is *not* science. And since within a modernist worldview, that which is unscientific is also assumed to be irrational, many modern scholars find it embarrassing and distasteful that such an important figure in the Scientific Revolution as Galileo was also an astrologer. Although James Robert Brown does not reference Galileo, he makes precisely this argument regarding astrology, stating that Popper's definition of science versus pseudoscience provides an impetus for scholars to see 'the flourishing of astrological and other such silly beliefs' as a 'great embarrassment to any evolutionary epistemologist'.[33] That attitude provides a strong motivation to ignore or explain away Galileo's astrological work rather than consider it within its historical context.

---

[28] Mario Biagioli, *Galileo, Courtier: The Practice of Science in the Culture of Absolutism* (Chicago: University of Chicago Press, 1994), Chapter 1.

[29] Richard DeWitt provides a useful overview of the concept of a worldview in *Worldviews* (Chichester: Blackwell, 2010, 2nd edition), section I.

[30] Castel and Sismundo, *The Art of Science*, p. 9.

[31] Alvin I. Goldman in his *Knowledge in a Social World* (Oxford: Oxford University Press, 1999), pp. 248–254, passim, states outright that to be rational is to be scientific. For Goldman, rationality is unchanging and invariably defined as following some approximation of the scientific method.

[32] Karl Popper, *Conjectures and Refutations* (London: Routledge and Keagan Paul, 1963), p. 34.

[33] James Robert Brown, *Smoke and Mirrors: How Science Reflects Reality* (London: Routledge, 1994), p. 62.

A major component of that context is to understand that Galileo's acceptance of predictive astrology was rational, when that term is properly understood. Even a cursory examination of the literature demonstrates that the meaning of such a seemingly basic and non-contentious word is rather more complex than we might initially assume. The debate about rationality, based as it is on such larger concepts as the nature of evidence and the ontological status of knowledge itself, is wide ranging and to attempt a solution here would take me too far afield from my study to be worthwhile. Nevertheless, a few clarifying comments are in order.

Steven Lukes argues that 'there are contextually-provided criteria for deciding what counts as a "good reason" for holding a belief'.[34] Therefore, 'rationality' is not an unchanging concept but is wholly dependent on the time and place under consideration. Indeed, as Peter Winch argues, there is no single rationality. Instead, this concept is only explicable on a contextual basis. Whether or not an individual is rational can only be adjudicated based upon whether or not he or she acts – or thinks – in a way that conforms to the norms of his or her culture.[35] This is not to suggest that the phenomenological world should be approached from a position of relativism. Rather, the focus of this approach is the way in which objective phenomena are interpreted and understood through the lens of the basic beliefs of the individual. Such beliefs are constructed upon the foundational knowledge and ideas the individual has acquired as part of his or her historical and cultural heritage, which creates a conceptual framework 'forced upon [the individual] by his experience of the world', as Richard Swinburne would put it.[36] Therefore, if we want to understand rationality in any given time and place, we must consider the background and cultural norms upon which a particular form of rationality is based. Just as Roy Willis and Patrick Curry argue that the nature of astrological *practice* is a reflection of its surrounding culture, so must beliefs as well as modes of thinking be placed within the context of the network or web of beliefs built up over time by intellectuals and other arbiters of cultural heritage in the society under consideration.[37]

---

[34] Steven Lukes, 'Some Problems About Rationality', *Archives of European Sociology* VIII (1967): pp. 247–264, p. 263.

[35] Peter Winch, 'Understanding a Primitive Society', *American Philosophical Quarterly* I (1964): pp. 307–324.

[36] Richard Swinburne, *Faith and Reason* (Oxford: Oxford University Press, 2005, 2nd edition), p. 17.

[37] For Curry, see Roy Willis and Patrick Curry, *Astrology Science and Culture, Pulling Down the Moon* (New York: Berg, 2004), pp. 77–86. For a discussion of

Within this model of rationality, rational actors are seen to *perform* rationality through 'a sort of metanoia [meaning a personal change or metamorphosis] marked in particular by a bracketing of beliefs and of ordinary modes of thought and language, which is the correlate of a tacit adherence to the stakes and the rules of the game'.[38] Well before Galileo's birth, the rules of the game he played were set by the arbiters of intellectual thought in Europe, who were primarily the masters of the various universities that had been established across Europe beginning with the University of Bologna in 1189 and in the field of astrology, heavily influenced by the writings of Arabic-language authorities on astrology.[39] Within this context scholars dealing with natural philosophical questions were almost always university graduates, who assumed that an investigation of the natural world through the application of rules established by Greek and Hellenistic thinkers such as Aristotle 384 BCE– 322 BCE) and Ptolemy (fl. 150 CE) represented what it was to be a rational individual.[40] The result was a premodern European intellectual class who shared a set of presuppositions taken for granted and held to be beyond dispute – a habitus in Pierre Bourdieu's terminology – inculcated through their common education.[41] The component parts of this education most relevant to our current study included the reading of a shared literature written primarily by Greek and Arabic authors and translated into Latin. Europe's intellectual elite then understood the ideas gleaned from this reading through shared forms of analyses driven by basic assumptions about the world derived from this common experience and reinforced by the authority of professors, degree-granting institutions, and the professionalization of the intellectual class in Europe.

This model of rationality is normative; the questions one asks and the basic contexts within which data are to be interpreted or ideas are

---

the concept of the web of belief, see Willard Van Orman Quine and J. S. Ullian, *The Web of Belief* (New York: Random House, 1978, 2nd edition).
[38] Pierre Bourdieu, 'The Peculiar History of Scientific Reason', *Sociological Forum* 6.1 (1991): pp. 3–26, p. 8.
[39] Edward Grant, *The Foundations of the Modern Sciences in the Middle Ages* (Cambridge: Cambridge University Press, 1996), Chapter 3. For astrology see Brady, pp. 77–100.
[40] For Bourdieu's analysis of the functioning of investigators within the modern scientific fields see Bourdieu, 'The Peculiar History of Scientific Reason', p. 8.
[41] Bourdieu, 'The Peculiar History of Scientific Reason', p. 8.

evaluated ordinarily derive from the society within which one lives.[42] Therefore, in the absence of revolutionary paradigmatic change, beliefs within a society typically develop from older ideas as individuals analyse the world around them through the lens of their education, experiences, religion, and other important aspects of their worldview. Very few if any beliefs can be considered freestanding; instead, concepts, interpretations, and analyses are conjoined to those a person has acquired through his or her cultural and intellectual heritage.[43] It is the conjunctive nature of beliefs that make astrology – in both its judicial and natural forms – in premodern Europe thoroughly rational, built as it was on an internally consistent set of principles validated both by empirical evidence interpreted through the lens of the astrological conceptual model as well as methods of analysis constructed over many centuries that had survived repeated challenge and testing. The history of the development of astrological systems of thought, from the fifth-century BCE? Babylonians who first cast horoscopes intended to describe an individual's future to the elaboration of Greek thinkers such as Plato (ca. 428–347 BCE) and Aristotle along with their Hellenistic successors like Ptolemy, as well as those of the Near East such as Abū Ma'shar al-Balkhī – known in the West as Albumasar (787–886) – all of which jointly created the system eventually bequeathed to Medieval Europe through the translation efforts of men such as Adelard of Bath (1080 –ca. 1152), is well covered elsewhere.[44]

However, in order to understand the rules of the game that Galileo played, we should know something of the intellectual edifice that had been built up by the time he received his education and began his work. In this way, we can see – at least in broad outlines – the worldview that Galileo worked within, in which astrological beliefs not only were rational, but in fact had a high measure of social capital, which the high status Aristotelian natural philosophy developed in the high Middle Ages only reinforced. Even though Galileo would do much to overturn Aristotelian physics, the

---

[42] Helen Longino, *Science as Social Knowledge* (Princeton: Princeton University Press, 1990), p. 12.

[43] Swinburne, *Faith and Reason*, pp. 4–9.

[44] For Babylonian horoscopes see Abraham Sachs, 'Babylonian Horoscopes', *Journal of Cuneiform Studies* VI (1952): pp. 54–56. For astrology in general see S. J. Tester, *A History of Western Astrology* (Woodbridge: Boydell Press, 1987) still provides a useful introduction. For a more thorough consideration, see Nicholas Campion, *A History of Western Astrology*, vols. I and II (London: Continuum, 2009).

Greek thinker both dominated Galileo's education and influenced his work.[45] Aristotle did not deal with astrology comprehensively, but what he had to say about celestial influence gave it an important place. In two key works translated into Latin between 1150 and 1160, the *De generatione et corruptione* (*On Generation and Corruption*) and the *Meteorologica* (*Meteorology*),[46] Aristotle argues that the Sun acts as the universal efficient cause as it moves around the ecliptic (its apparent path on the celestial sphere), making it necessary for both generation and corruption.[47] In Aristotelian terms, that means the influence of the Sun is necessary in the sublunar realm for both coming-to-be and passing away, which includes both being born and dying for living things. In other words, as the efficient cause, the Sun was equally important to the process of conception and birth as the male – who provided the formal cause of the seed – and the female – who provided the material cause in the womb. Although Aristotle did not discuss the influence of other celestial bodies, neither did he directly rule such influence out. Therefore, the door was wide open for later writers working broadly from an Aristotelian position to articulate a conception of heavenly influences upon terrestrial creatures that was both far more complex and more specific than Aristotle's general notion that the Sun is essential for coming-to-be and passing away in the sublunary realm.

While Aristotle may have established some of the important preconditions for those who would later evidence an interest in astrology – especially in the Latin Christian West – most Greek writing on astrology

---

[45] William A. Wallace, *Galileo, the Jesuits, and the Medieval Aristotle* (Farnham: Ashgate, 1991), Chapter 1.

[46] John D. North, 'Medieval Concepts of Celestial Influence: A Survey', in *Astrology, Science and Society. Historical Essays*, ed. Patrick Curry (Boydell Press: Woodbridge, 1987): pp. 5–18, p. 5; John North, 'Celestial Influence—the Major Premiss of Astrology', in *'Astrologi hallucinati': Stars and the End of the World in Luther's Time*, ed. Paola Zambelli (New York: Walter de Gruyter, 1986): pp. 45–100; Amable Jourdain, *Recherches Critiques sur l'Âge et l'Origine des Traductions Latines d'Aristote et sur des Commentaires Grecs ou Arabes employés par les Docteurs Scolastiques* (New York: Burt Franklin, 1960, 2nd edition), pp. 31, 37, 75, 124, 173, 327.

[47] Aristotle discusses this in *De generatione et corruption*, ed. Harold Joachim (Oxford: Clarendon Press, 1922), 2.10. A useful summary of his views may be found in Gad Freudenthal, 'The Astrologization of the Aristotelian Cosmos: Celestial Influences on the Sublunary World in Aristotle, Alexander of Aphrodisias, and Averroes', *New Perspectives on Aristotle's De caelo*, ed. Alan C. Bowen and Christian Wildberg (Leiden: Brill, 2009): pp. 239–281. I rely on this source for the rest of the information in this paragraph.

would come in the centuries after his death.[48] None of these later writings would be as important as the second century CE Alexandrian, Ptolemy. For the purposes of this study, his most important works are the *Almagest*[49] and the *Tetrabiblos*.   The first of these is primarily an astronomical work detailing a geocentric model of the cosmos, while the latter includes the basis for a model of judicial astrology. However, Ptolemy did not separate the disciplines as a modern person might, and in fact we should be very careful in making distinctions between astrology and astronomy when discussing the premodern study of the heavens. First, the terms themselves are not clearly delineated until the modern period, and in fact variations of 'astronomy' are often used to indicate a discipline allowing prediction of the future.[50] Second, Ptolemy drew a very different distinction than modern people would: he established a division between two different predictive approaches to studying the stars without distinguishing the terms used for those different approaches. Instead, his distinction was between approaches that provide 'theoretical' and certain knowledge.[51]

For Ptolemy, the study of the motions of the heavens, including the predictions possible about those motions, functions at a level of mathematical certainty that no one could doubt.[52] After all, the heavens are perfect and unchanging, so how could they not behave with absolute regularity? However, any discipline dealing with the changeable realm of elementary bodies in the sublunar world must suffer from the unpredictability found in such bodies. Therefore, mathematical analyses of heavenly bodies interacting with one another are certain while any analysis of how the terrestrial realm affects the sublunary world provides only possible knowledge. Nevertheless, so long as the judicial astrologer is properly trained in the mathematical arts, is careful, and does not promise

---

[48] Tamsyn Barton, *Ancient Astrology* (London: Routledge, 1994), Chapter 2.

[49] This title means simple 'The Greatest'. A discussion of the process of transmission can be found in 'Almagest: its Reception and Influence in the Islamic World', in *Encyclopedia of the History of Science, Technology, and Medicine in Non-Western Cultures*, ed. Helaine Selin (Dordrecht: Kluwer, 1997), pp. 55–56.

[50] For a discussion of the changing uses of this term, see Scott E. Hendrix, *How Albert the Great's* Speculum Astronomiae *Was Interpreted and Used by Four Centuries of Readers: A Study in Late Medieval Medicine, Astronomy and Astrology* (Lewiston: Edwin Mellen, 2010), Chapter 6.

[51] The best overview of the *Almagest* can be found in Olaf Pederson, *A Survey of the Almagest* (New York: Springer, 2010).

[52] Pederson, *A Survey of the Almagest*, p. 401.

too much, the discipline can be very useful.[53] Ptolemy's work would remain the basis for astronomical education well into the early modern period; Galileo quoted the *Almagest* in his own notes to a course on cosmography he taught for the first time in 1602.[54]

In the twelfth century, translators such as Adelard of Bath began to reintroduce Greek and Hellenistic natural philosophy to the West; meanwhile a surge in translation of Arabic works applied Aristotelian principles with Neoplatonic accretions to an exposition of astrological theory and practice that was complex both from a standpoint of its technical sophistication as well as its philosophical justifications and conveniently worked out within a monotheistic context.[55] Armed with the tools provided by Arabic writers such as Albumasar, whom Brady has argued was of particular importance to Galileo's later work,[56] European interest in this celestial discipline exploded. Very quickly, physicians integrated astrology into their treatments and the European nobility began to employ astrologers as advisers,[57] causing the discipline to grow in profile.

---

[53] Pederson, *A Survey of the Almagest.* p. 402.

[54] Stillman Drake, *Galileo at Work: His Scientific Biography* (Chicago: University of Chicago Press, 1978), pp. 51–53.

[55] Richard Lemay, *Abu Ma'shar and Latin Aristotelianism* (Beirut: The Catholic Press, 1962), Introduction and Chapter 1; Richard Lemay, 'The True Place of Astrology in Medieval Science and Philosophy: Towards a Definition', in *Astrology, Science and Society: Historical Essays*, ed. Patrick Curry (Woodbridge: Boydell Press, 1987): pp. 57–73, 65–68; Scott E. Hendrix, 'Reading the Future and Freeing the Will: Astrology of the Arabic World and Albertus Magnus', *Hortulus* 1 (2006): pp. 30–49.

[56] Brady, 'Galileo's Astrological Philosophy', pp. 77–100.

[57] The medical school at Padua immediately embraced Aristotle's *libri naturales* as well as the works of Albumasar, turning out physicians who were well versed in astrology. See Ferdinand Van Steenberghen, *Aristotle in the West* (Louvain: Nauwelaerts, 1955), pp. 62–66. For one of Padua's more notorious physician and astrologers, Guido Bonatus (ca. 1210–ca.1300), see George Sarton, *Introduction to the History of Science* (Baltimore: Williams and Wilkins, 1931; reprinted 1961), Vol. II, pp. 988–989. Careers combining astrological forecasting and advising with the practice of medicine were common. See Nancy Siraisi, *Medieval and Early Renaissance Medicine* (Chicago: University of Chicago Press, 1990), p. 68. Frederick II was one of the earliest rulers to employ an astrologer, and he in fact employed two: Michael Scot and Master Theodore. See Hilary Carey, *Courting Disaster: Astrology at the English Court and University in the Later Middle Ages* (New York: St. Martin's Press, 1992), p. 31.

Nevertheless, this growing integration of judicial astrology into the European intellectual milieu was anything but smooth. Early Christian writers such as Origen (185-255) and Augustine (354-430) mistrusted astrology,[58] but the key point of contention was concern that the ability to make predictions about the future would negate human free will, thereby making God, not humankind, responsible for sin.[59] Nevertheless, even such critics as Augustine accepted the principles of natural astrology, at least in its broad outlines, and for centuries there was little potential for a clash between judicial astrology's supporters and detractors. The disappearance of the Roman educational system in the sixth century ended the study of learned astrology until the thirteenth century, muting the controversy over the discipline.[60]

Controversy about judicial astrology re-emerged in the high Middle Ages in ways that are important to understand if we are to understand how astrology fit into the well-articulated understanding of the cosmos that Galileo inherited. The key events occurred in the late thirteenth century in a complex series of conflicts and coalescence of attitudes that makes understanding the controversial nature of astrology in the high Medieval period no easy task. A generalized suspicion of Aristotle as a pagan was

---

[58] Laura Ackerman Smoller, *History, Prophecy, and the Stars: The Christian Astrology of Pierre d'Ailly, 1350-1420* (Princeton: Princeton University Press, 1994), pp. 26–27; Theodore Otto Wedel, *The Mediaeval Attitude Towards Astrology, particularly in England* (New Haven: Yale University Press, 1920), pp. 11–12.

[59] Lynn Thorndike, *History of Magic and Experimental Science* (New York: Columbia University Press, 1923), Vol. I, pp. 456–458; Barton, *Ancient Astrology*, p. 75; Tim Hegedus, *Early Christianity and Ancient Astrology* (New York: Peter Lang, 2007), pp. 51–58, 162, 329–334.

[60] Valerie Flint has demonstrated that astrology never lost its fascination for Europe's dwindling numbers of educated men and women in the early middle ages. See Flint, *The Rise of Magic in Early Modern Europe* (Princeton: Princeton University Press, 1991), pp. 93, 99. However, without the tools of the astrologer's trade – primarily Greek texts and tables drawn up for the location of any given horoscope – such interest would have been uninformed and any *practice* of astrology would have been impossible. See Carey, *Courting Disaster*, p. 27; Smoller, *History, Prophecy, and the Stars*, p. 29. True, the Councils of Toledo (400) and Braga (560–565) condemned astrology. However, M. L. W. Laistner argues that these condemnations were aimed at the Priscillianists, for whom belief in astrology constituted a religious dogma. See Laistner's 'The Western Church and Astrology during the Early Middle Ages', *Harvard Theological Review* 34 (1941): pp. 251–275, 264, 275.

partly responsible, but interdisciplinary turf wars were more important. As Greek and Arabic ideas became more familiar to the intelligentsia of Europe, arts masters in the new universities sought greater recognition by applying Aristotelian philosophy to topics that seemed to intrude upon the province of their more prominent colleagues in the theology faculty. The combination of factors means it is not always easy to discern when opposition was truly aimed at astrological models of the world and when this stance was merely a convenient excuse for the airing of larger grievances.[61] Regardless of the exact reasons behind the opposition, by the late thirteenth century astrological doctrines prompted vigorous attacks.

In this atmosphere of acrimony and suspicion, the two greatest scholars of the day, Albert the Great (1206–1280) and his student Thomas Aquinas (1225–1274), articulated a defence of astrology that became central to the rational worldview Galileo inherited. Of the two, Albert is the more relevant to the current topic. As William A. Wallace has demonstrated, Albert had a direct influence on Galileo, which is why I will focus my attention on him.[62] Albert's influence on Galileo's thought is all the more interesting since Albert had an intense interest in astrology, with astrological references finding their way into almost everything he wrote. The breadth and depth of his interest in the subject means that a full analysis of his position would take us too far afield from our current topic.[63] For now, let us consider the core of Albert's defence of the discipline. He maintained that the human soul was necessarily of a higher order of substance than the body.[64] This statement was undoubtedly influenced by the strains of Neoplatonism that are everywhere evident in

---

[61] Van Steenberghen, *Aristotle in the West*, pp. 69–70; Gordon Leff, *Paris and Oxford Universities in the Thirteenth and Fourteenth Centuries* (New York: John Wiley & Sons, 1968), pp. 193–197.

[62] William A. Wallace, 'Galileo and Albertus Magnus', *Boston Studies in the Philosophy of Science* 62 (1981): pp. 264–285.

[63] For that, see Hendrix, *Albert the Great's* Speculum astronomiae, Chapter 3. For a detailed discussion of Albert's view of astrology and its relationship to free will, see Scott E. Hendrix, 'Choosing to be Human: Albert the Great on Human Agency and Celestial Influence', *Culture and Cosmos* 12, no. 2 (2008): pp. 23–41.

[64] Albert the Great, *De causis et processu universitatis a prima causa II: Opera omnia,* ed. Winfrid Fauser (Monasterii Westfalorum: Aschendorff, 1993), Vol. XVII, p. 57; Albert the Great, *De caelo et mundo*, ed. Paul Hossfeld (Monasterii Westfalorum: Aschendorff, 1971), p. 114; Albert the Great, *Liber de natura et origine animae: Opera omnia,* ed. Bernhard Geyer (Monasterii Westfalorum: Aschendorff, 1971), Vol. XII, p. 12.

Albert's thought, drawn largely from his mistaken belief that the *Liber de causis* (*Book about the Causes*) was a section of Aristotle's *Metaphysics* rather than a paraphrasing of ideas drawn from Proclus and Avicebron.[65] Because of this soul/body distinction the stars as corporeal bodies were held to influence the body directly, but the soul *per accidens*, indirectly.[66] Therefore, the will, which is a component of the intellectual soul, is free to resist corporeal impulses imparted by the stars. To explain this idea Albert cites the maxim, 'the wise man will dominate the stars,' a rationale drawn directly from Albumasar's *Introductorium maius* (*Greater Introduction*).[67] Albert elaborates that one learned in the influences of the heavens can avert many negative things, while maximizing positive effects – if one only makes the willed effort to do so.[68] Unfortunately, most people rarely exercise their will to oppose corporeal impulses, which means that astrological predictions are usually accurate if performed correctly. In this way Albert outlines a model of celestial influence that allows for judicial astrology without compromising the freedom of the will. This point is of enormous import, for it cleared a space for judicial astrology in the Latin

[65] Alain De Libera, *Albert le Grand et la Philosophie* (Paris: J. Vrin, 1990), pp. 55–59. This mistake was universal prior to Thomas of Moerbeke's completion of a new translation directly from the Greek in 1268. See Ferdinand Van Steenberghen, *The Philosophical Movement in the Thirteenth Century* (Edinburgh: Thomas Nelson and Sons, 1955), p. 40.

[66] Albert the Great, *Speculum astronomiae,* in *The Speculum Astronomiae and its Enigma: Astrology, Theology, and Science in Albertus Magnus and his Contemporaries*, ed. Paola Zambelli (Dordrecht: Kluwer Academic Publishers, 1992), p. 220, pp. 250–256, Chapters 3 and 12. I am aware that there is much resistance to the idea that Albert wrote the work now known as the *Speculum astronomiae*. For a full discussion of the controversy and why I accept the attribution of Albert's authorship of this work as genuine, see Hendrix, *Albert the Great's* Speculum astronomiae, Chapter 1. For a thorough examination of the alternative position – which I will address in a future article – see Jeremiah Hackett, 'Albert the Great and the *Speculum astronomiae*: The State of the Research at the Beginning of the 21st Century,' in *A Companion to Albert the Great*, ed. Irven M. Resnick (Leiden: Brill, 2013), pp. 437–450.

[67] For Albumasar as the source of this maxim in the West, see Lemay, *Abu Ma'shar*, pp. 42–48. G. W. Coopland attempts to trace the provenance of this maxim in appendix four of his work, *Nicole Oresme and the Astrologers: A Study of the Livre de Divinacion* (Cambridge, MA: Harvard University Press, 1952), pp. 175–177.

[68] Albert, *Speculum*, pp. 258–261, Chapter 13. This is also a central point in his shorter, and less-well-known, *De fato*.

Christian West. This remained the prevalent view well into the seventeenth century, and it is precisely what Galileo would have learned about astrology.

Albert's star student, Thomas Aquinas, steadfastly supported Albert's view of astrology even if the issue was not a primary concern for the younger Dominican. Thomas accepted that the stars as corporeal bodies influenced human bodies directly, but affected the incorporeal soul only indirectly through the sensory powers that produce 'phantasms' in the intellect.[69] In this way, celestial influence might 'incline' a person toward a certain action, but since it is always possible to resist such inclinations through an exercise of the will, the stars impart no necessity upon human action.[70] Reiterating the dictum passed on by his master, Thomas notes that the 'wise man is master of the stars', but since most people are ruled by their passions astrological predictions are normally efficacious.[71] Again, this point is key and represents the dominant view of the educated elite in Europe down to the time of Galileo and beyond: astrological predictions can provide *probable* predictions about the future. Any given prediction involving human beings may prove to be wrong, if the person or people in question choose to exercise their will to negate astral influences. However, since most people rarely do so, most predictions will come to pass if the astrologer performs his task well. Thomas accepted this view and saw little reason to discuss astrology at any length, even if he did write about it of it in some one hundred and thirty passages, as compiled by Thomas Litt.[72]

---

[69] Thomas Aquinas, *Summa theologiae* (Cambridge: Blackfriars, 1964), IIa IIae q. 95 a. 5; Ia q. 115 a. 3; John R. Bowlin, *Contingency and Fortune in Aquinas's Ethics* (Cambridge: Cambridge University Press, 1999), p. 59; Norman Kretzmann, 'Philosophy of Mind,' *The Cambridge Companion to Thomas Aquinas*, ed. Norman Kretzmann and Eleonore Stump (Cambridge: Cambridge University Press, 1993), p. 148.

[70] Thomas, *Summa*, Ia, q. 115, a.3. For a full analysis of Thomas' view of astrology and its relationship to free will, see Scott E. Hendrix, 'Contemplating the Stars and Comprehending Humanitas', pp. 11–22.

[71] Thomas, *Summa*, Ia, q. 115, a.3: 'ipsi astrologi dicunt quod sapiens homo dominatur astris... plures hominum sequuntur passiones... Et ideo astrologi ut in pluribus vera possunt praedicare, et maxime in communi'. In English: 'The astrologers themselves state that the wise man dominates the stars [however] many men follow the passions... therefore, astrologers are able to predict true things in many instances, and especially in common [circumstances].' All translations are my own.

[72] Thomas Litt, *Les corps célestes dans l'univers de Saint Thomas d'Aquin* (Paris: Publications Universitaires, 1963).

Thomas's writings on astrology were nonetheless important for the promotion of Albert's definition of 'superstitious' versus licit astrology, a view that Thomas presented with even greater clarity than had his master. As Laura Ackerman Smoller has pointed out, Thomas expressed concern that attempts to predict the future with too great an accuracy or specificity could lead men to mingle with demons, thus indulging in 'superstitious' forms of astrology.[73] Nevertheless, he allowed that one might predict general events caused by celestial influence, such as weather patterns.[74] In this manner, Thomas sought to preserve a worldview derived from Aristotelian physics and cosmology that took celestial influence as a given[75] while leaving the door open for a form of judicial astrology that did not compromise the Christian faith.[76] Given the weight of Thomas' reputation, especially after his canonization in 1323, his arguments about the permissibility of certain forms of astrology would prove to be quite influential. After all, few or none denied the central premise of natural astrology, which also provided the basis for the judicial application of the discipline, that heavenly bodies influenced terrestrial creatures.[77]

Albert's and Thomas' views about astrology are important if we are to understand the worldview informing Galileo's beliefs on the subject. Their arguments became normative during the late Medieval period and would remain so well into the early Modern period. In spite of recurrent attacks on the study or practice of astrology, such as we find at Paris in 1270 and 1277, writers would again and again reference Albert and sometimes Thomas when discussing their own views about astrology, particularly as the popularity of the subject grew in the closing years of the thirteenth

---

[73] Smoller, *History, Prophecy, and the Stars*, p. 31; Thomas, *Summa*, IIa IIae q. 95, a. 5.
[74] Smoller, *History, Prophecy, and the Stars*, p. 31; Thomas, *Summa*, IIa IIae q. 95, a. 5.
[75] This was, after all, the unifying theory that infused the medieval view of the cosmos. See North, 'Celestial Influence', pp. 45–100.
[76] Tester, *A History of Western Astrology*, pp. 182–183.
[77] Stefano Caroti, 'Nicole Oresme's Polemic Against Astrology, in his *Quodlibeta*', *Astrology, Science and Society*, ed. Patrick Curry (Woodbridge: The Boydell Press, 1987), pp. 75–93, p. 78. Even zealots who opposed astrological forecasting, such as Bernardino da Siena, did not question this idea. See Garin, *Astrology in the Renaissance*, pp. 31–32. I use the term 'creature' in its original meaning, as that which was created by God, including animate as well as inanimate objects.

century.[78] Nowhere was this growth in popularity more apparent than in the medical field. By the late thirteenth century, physicians such as Pietro d'Abano (ca. 1250–1318) began aggressively to defend the use of astrology in medicine[79] as vital both to diagnoses as well as to treatments.[80] This position became well entrenched among both physicians and surgeons, with Marsilio Ficino (1433–1499) being a strong proponent of medical astrology in the fifteenth century,[81] a tradition maintained by Giambattista della Porta (ca. 1537–1615) well into the early-modern period.[82] These medical professionals saw astrology as a vital diagnostic tool; from the thirteenth through the seventeenth century no reputable physician or surgeon would dream of rendering a diagnosis or applying a treatment without first considering the role of celestial influences, and we should remember that Galileo initially trained as a physician.[83] Beginning

---

[78] I am referring to the condemnations of 1270 and 1277, the latter (and more famous of which) may be found in Henri Denifle and Emile Chatelain, O.P., eds, *Chartularium Universitatis Parisiensis* (Paris: Delalain, 1889), Vol. I, p. 541. The factors leading to, and the impact of, these condemnations are quite complicated. See Etienne Gilson, *History of Christian Philosophy in the Middle Ages* (New York: Random House, 1955), pp. 405–406; Leff, *Paris and Oxford Universities*, pp. 231–238; See John F. Wippel, 'The Condemnations of 1270 and 1277 at Paris', *The Journal of Medieval and Renaissance Studies* 7, no. 2 (1977): pp. 169–201, p. 190; Roland Hissette, *Enquête sur les 219 articles condamnés à Paris le 7 mars 1277* (Louvain: Publications Universitaires, 1977).

[79] Graziella Fredirici Vescovini, 'Peter of Abano and Astrology', in *Astrology, Science and Society*, ed. Patrick Curry (Woodbridge: The Boydell Press, 1987): pp. 19–40, pp. 20–21.

[80] Pietro d'Abano, *Conciliator Controversiarum quae inter philosophos et medicos versantur* (Mantua: Ludovicho de Gonzaga, 1477), ff. 20v-20r. Pietro states, 'qui diligenter inspiciunt concedunt hanc scientiam astronomiae non solum utilem sed et necessariam maxime medicinae', or in English: 'those who diligently consider this science of astrology concede it to be not only useful, but also necessary, especially to medicine'. f. 22r. See also Vescovini, 'Peter of Abano and Astrology', pp. 19–40, pp. 20–21.

[81] Don Cameron Allen, *The Star-Crossed Renaissance* (Durham: Duke University Press, 1941), p. 8. Ficino developed numerous pharmaceutical recipes that included directions for admixture and administration according to astrologically propitious times; he also advocated the use of astrological images as a form of medical treatment.

[82] Io. Baptista Porta, *Magiae naturalis libri viginti* (Rouen: Ioannis Berthelin, 1650), Vol. I, ff. 2v-2r.

[83] Roger French, *Medicine Before Science* (Cambridge: Cambridge University Press, 2003), pp. 130–135.

in the early fourteenth century when Aristotelian natural philosophy was still rather new and exciting to Italian physicians,[84] these professionals appealed to Aristotle's *libri naturales* for an understanding of physiology – sometimes even in preference to Galen or other medical authorities.[85] This reliance upon Aristotle naturally enhanced the appeal of astrology, given the philosopher's support of the theory of celestial influence found in his *De generatione et corruptione* and *Meteorologica*.[86] For these physicians, since celestial motion was assumed to provide the motive force for conception, it logically followed that the movements of the heavens brought about changes in health throughout the course of a person's life.[87] By extension, these celestial influences affected people's lives in all sorts of ways, which is why those practicing judicial astrology were often physicians who applied their education to the casting of nativities (predictions about a person based on his or her time of birth), elections (a form of astrology presumed to highlight the most propitious time for an undertaking), or a host of other prestigious and lucrative uses of astrology.

The important point is that European intellectuals widely accepted the idea that the heavens influenced terrestrial creatures and events in a wide variety of ways. Even those most sceptical of astrology could agree on that. Consider the case of Pierre d'Ailly (1350–1420). Laura Smoller has detailed the way in which scepticism of judicial astrology in d'Ailly's early work eventually transformed itself to an interest in celestial divination as a tool to predict the end times, providing an alternative to the proliferation of uncontrolled prophets who arose during the Great Schism.[88] However, it is clear that even at his most sceptical, d'Ailly never rejected the idea that humankind and the rest of terrestrial creation exist at the centre of a web of celestial influences, greatly affecting all sublunar creatures,[89] which he accepted in part due to the influence of Albert the Great.[90]

---

[84] Nancy Siraisi, *Taddeo Alderotti and his Pupils. Two Generations of Italian Medical Learning* (Princeton: Princeton University Press, 1981). Siraisi suggests that Taddeo may have introduced Aristotelian natural philosophy into Italian medical learning in the 1260s.

[85] Nancy Siraisi, 'Taddeo Alderotti and Bartolomeo de Varignana on the Nature of Medical Learning', *Isis* 68, no. 1 (1977): pp. 27–39, p. 29.

[86] North, 'Medieval Concepts of Celestial Influence', p. 5.

[87] Siraisi, *Taddeo Alderotti*, p. 179.

[88] Siraisi, *Taddeo Alderotti*, Chapters three and five.

[89] Smoller, *History, Prophecy, and the Stars*, pp. 43–45.

[90] Pierre d'Ailly, *Apologia defensiva astronomiae ad magistrum Johannem cancellerium parisiensem* (Louvain: J. de Paderborn, 1483), f. 143v:

Not all agreed with D'Ailly, of course. His student, Jean Gerson (1363–1429), would never accept his master and old friend's arguments in favour of astrological divination. Gerson does appear to have agreed with the central arguments in favour of celestial influence found both in the later work of Pierre d'Ailly as well as his predecessor, Albert the Great,[91] as we can see in his *Tricelogium astrologiae theologizatae* (*The Tripartite of Theologized Astrology*). There he writes: 'I admit that heaven works strongly upon or influences such things that have been begun',[92] so long as one is mindful of the standard caveat, that such influence 'induces no necessity whatsoever into men, but only an inclination'.[93] Nevertheless, Gerson argued that after the birth of Christ it became impossible to predict future events. The new order that Christ ushered in was one in which the actions of humankind, based as they are on free will, were so varied and unpredictable that it is beyond human capacity to foresee what may come to pass with any measurable degree of accuracy, therefore negating the standing of astrology as an *ars*.[94] Moreover, the *Tricelogium* presents a

---

'Concordemus denique cum Alberto magno doctore sancti Thomae in illo praecipue tractatu suo qui Speculum dicitur, ubi hanc materiam plene utiliterque pertractat', or in English: 'We agree, then, with Albert, the great teacher of the sainted Thomas, especially in his tract which is called the Speculum, where he thoroughly, clearly, and usefully deals with this matter [of astrology].'

[91] Jean Gerson, *Tricelogium astrologiae theologizatae*, in his *Oeuvres Complètes*, ed. Mgr. P. Glorieux (Paris: Desclée, 1962), p. 96.

[92] Gerson, *Oeuvres Complètes*, caput X, p. 111. 'Admisso quod caelum in talibus initiis fortius agit aut influit.'

[93] Gerson, *Oeuvres Complètes*, caput X, p. 112. 'Nihilominus necessitatem in hominibus nullam, sed tantummmodo inclinationem adducit.'

[94] Gerson, *Oeuvres Complètes*, caput X, pp. 110–111. 'Hanc vero artem [astrology] vel principia eius probare volumus semper extitisse extra et supra totam humanae investigationis facultatem. Ars quippe certa et regularis esse debet; voluntates autem hominum et cogitationes secundum quas deberet talis ars judicativa, capere fundamentum, penitus incertae sunt et variae; ut ergo nulla sit naturaliter ars de eis constantissime fatendum est. Numquid advertimus post Christi Nativitatem (quae utique miraculosa fuit, nec influentiis subjecta,) quanta varietas in conditionibus, moribus et operibus hominum innumerabilium secuta est' or in English : 'But I wish to prove that this art (astrology) or its principles has always been beyond every ability of human investigation. An art should, by all means, be certain and regular ; such a judicial art should be able to grasp the foundational desires and thoughts of men according to its principles, but they [human desires and thoughts] are deeply uncertain and variable ; therefore, the art naturally can say nothing about those things with great accuracy. Surely we should

distinctly suspicious attitude toward the study of astrology in its various forms, as having become contaminated with so many superstitions since the time of the patriarchs that its study threatened to drag the soul of anyone unwise enough to undertake it down into eternal darkness.[95] Additionally, the study of astrology encourages men to focus upon the mediating matter of the heavens, rather than the majesty of God, thereby giving birth to idolatry.[96]

---

not pay attention [to astrological judgments] after the Nativity of Christ (which in any case was miraculous, not subject to influence), so great are the variations that are followed, in conditions, customs, and works, of the uncountable number of men.' Gerson, propositio X of his *Tricelogium*, p. 96. Also see Jean Gerson, 'De respectu coelestium siderum,' *Oeuvres Complètes*, Mgr. P. Glorieux, ed. (Paris: Desclée, 1962), vol. X, pp. 109–116. In particular note the scathing rebuke to those 'homines idiota et simplices' who put faith in such things. My thanks to Drs Jeffrey Fisher and Kevin Guilfoy for discussing Gerson's Latin, and the intent behind it, with me. Jeffrey Fisher's work on Gerson should be the starting point for anyone interested in understanding the noted conciliarists thought. See Jeffrey Fisher, 'Jean Gerson', in *The Late Medieval Age of Crisis and Renewal, 1300-1500: A Biographical Dictionary*, ed. Clayton J. Drees (London: Greenwood Press, 2001): pp. 183–185; Jeffrey Fisher, 'Gerson's Mystical Theology', in *A Companion to Jean Gerson*, ed. Bernard McGuire (Leiden: E.J. Brill, 2006): pp. 205–248.

[95] Gerson, *Tricelogium*, p. 90, preface: 'Propterea non est negandum ab astrologia, quam esse sciantiam nobilem et admirabilem primo patriarchae Adam et sequacibus revelatam theologia non abnegat. Verumtamen hanc ancillam suam astrologiam nonnulli tot vanis observationibus, tot impiis erroribus, tot superstitionibus sacrilegis deturpantes maculaverunt, nescientes in ea sobrie sapere et modeste uti, quod apud bonos et graves redita est necdum infamis sed religioni christianorum suisque cultoribus pestilens et nociva.' In English : 'Therefore, it must not be denied about astrology, that it was a noble and admirable science for the first patriarch Adam, and for those who followed, and theology will not deny that. But even so, so many defilers have polluted this handmaiden astrology with so many vain observations, impious errors, and completely superstitious sacrileges, to attempt to use it soberly, wisely, and modestly, it is such that among good and serious men, that which is not just notorious but also pestilential and noxious is returned to the religion and worship of these Christians [who make use of astrology].'

[96] Brian P. Levack, *The Witch-Hunt in Early Modern Europe* (New York: Longman, 1995, 2nd edition), p. 34. Concerns that individuals focusing too much on the created world would be led into idolatry similarly motivated Gerson to promote the denunciation of all magicians, white or black, as idolaters at Paris in 1398.

However, it is again important to note that Gerson accepts the idea that celestial influences 'incline' men toward certain actions and behaviours. The reason why he rejects astrological divination is because of the primacy of free will, the complicated nature of the various influences, and his fear that the study of astrology may promote idolatry. However, even this harsh critic of the use of astrology accepted Albert and Thomas' core premise – that celestial influence affects terrestrial creatures, making astrological predictions at least theoretically possible. In this, Gerson was simply speaking as a member of the learned community of his day. A belief that humanity sits at the centre of a complex web of celestial influence represented something of a grand unifying theory, welding together fields of knowledge as seemingly disparate as physics, metaphysics, and theology, all through the dominant philosophical school of the day: Aristotelianism.[97] Astrology was not simply a part of the intellectual landscape, but was rather a network of fibres running throughout this landscape, weaving together its constituent parts by explaining humankind's place within God's creation while allowing those knowledgeable of astrology's secrets to learn something of God and his plans through an analysis of his work and the influence he imparted to it.

This view of the cosmos, as one in which celestial influence affects terrestrial objects in ways that make predictions of future influences simply a matter of mathematical astronomy, was normative through the Renaissance and into the early modern period. Even those who have been held to reject this notion, when read carefully, offer a more complicated picture. For example, consider Giovanni Pico della Mirandola (1463–1494). As the author of a work known as the *Disputationes adversus astrologiam divinatricem* (*Disputations Against Astrological Divination*), it makes sense that scholars such as Eugenio Garin normally read him as a late-life convert to the rejection of astrology. After all, though this work presents many challenges to the reader, the title at least appears straightforward enough – though more on that in a moment. However, it was unfinished when Pico died, and certainly lacks the polish that Pico would have undoubtedly provided had he lived long enough to edit it.

---

[97] See John North's 'Medieval Concepts of Celestial Influence', pp. 5–18, and 'Celestial Influence', pp. 243–300, as well as Stefano Caroti, 'Nicole Oresme's Polemic Against Astrology', pp. 75–93, p. 78. In the words of Caroti: 'with varying degrees of emphasis, this influence [of the heavens over sublunary creatures] had come to be unanimously accepted by the Aristotelian scholastic tradition—to such an extent that it had become a topos in certain areas of commentaries on Aristotle's works.'

Furthermore, Luca Bellanti suggested as early as 1502 that the views it contain might be more those of Girolamo Savonarola (1452–1498) rather than Pico.[98] Nevertheless, modern scholarship has tended to portray the *Disputationes* as a philosophical assault on astrology born out of a rationally motivated rejection of 'superstition'.[99] Unfortunately, this interpretation of Pico's writing fails to take into account his expressed views on astrology presented in his other works and does not appear to be reflective of a close reading of the *Disputationes* themselves.[100] While Pico did become alarmed late in his life by certain forms of celestial divination, his critique of astrology in the *Disputationes* is not the wholesale rejection of the discipline that so many have taken it to be. The title of the work is not even his – his nephew, Giovanni Francesco Pico della Mirandola, edited the work after Pico's death, and he also supplied the title.

A complete re-evaluation of Pico's attitudes toward astrology is too far outside the boundaries of my present focus to merit inclusion here.[101] Yes, he does have harsh words for astrology 'prohibited by law, damned by the prophets, ridiculed by saints, forbidden by popes and sacrosanct synods'.[102]

---

[98] Allen, *The Star Crossed Renaissance*, p. 35. See also G. C. Garfagnini, 'Pico e Savonarola,' in *Pico, Poliziano e l'Umanesimo di fine Quattrocento,* P. Viti, ed. (Florence: L.S. Olshki, 1994): pp. 149–157; William G. Craven, *Giovanni Pico della Mirandola, Symbol of His Age: Modern Interpretations of a Renaissance Philosopher (Geneva: Libraire Droz, 1981).*

[99] For example, see Garin, *Astrology in the Renaissance*, pp. 87–93. Garin's misunderstanding of Pico may be coloured by incorrect assumptions about medieval and early-modern astrology. Garin seems to perceive astrology as involving some form of mind reading (pp. 35–37) and, in his description of the *Speculum* states that astrological theory predicated celestial influence only over the birth of an individual (p. 38), which is far from the case.

[100] H. Darrel Rutkin has examined the *Disputationes* and the contexts surrounding its production in 'Astrology, Natural Philosophy and the History of Science, c. 1250-1700: Studies Toward an Interpretation of Giovanni Pico della Mirandola's *Disputationes adversus astroligiam divinatricem* (unpublished PhD dissertation, Indiana University, 2002).

[101] For a start in that direction, see Scott E. Hendrix, 'Rational Astrology and Scientific Rationalism in Premodern Europe', in *Rational Magic*, ed. Scott E. Hendrix and Brian Feltham (Oxford: Fisher Imprints, 2011): pp. 65–105.

[102] Giovanni Pico della Mirandola, *Disputationes Adversus Astrologiam Divinatricem*, ed. Eugenio Garin. (Florence: Vallecchi Editore, 1946), Vol. 1, p. 94. 'Quis iam igitur audeat homo christianus (cunctis enim nunc mihi sermo) astrologiam tueri, sequi, extolerre, a lege prohibitam, a prophetis damnatum, a sanctis irrisam, a pontificibus et sacrosanctis synodis interdictam?' In English:

But he accepted that the heavens transmitted influence to terrestrial creatures, including people, stating 'we defend this [belief in celestial influence] as far as this, that nothing comes to us from heaven except with light having carried it',[103] and he predictably references Albert the Great on this point. He does castigate 'casters of nativities' as promoting 'the most infectious of all frauds',[104] while calling into question the system of affinities and antipathies so important to astrology and warning that this part of the discipline could lead the unwary into superstition.[105] However, we should be mindful again that harsh statements about fraudulent uses of astrology were a common staple among astrologers at this time as well. Pico goes on to caution against assigning too much strength to celestial influence, which is transmitted from a distant source by means of a vehicle – light – that is easily blocked.[106] However, one will look in vain for a clear statement that astrological forecasting is impossible.[107] Rather, Pico seems

---

'Therefore, what Christian man should now dare to examine astrology, follow, or extol astrology (with all these now [considered] in my discussion), prohibited by law, damned by the prophets, ridiculed by the saints, and forbidden by popes and sacred synods?'

[103] Pico, *Disputationes*, Vol. I, p. 253. 'quod hactenus defendamus, nihil ad nos a caelo nisi luce vehente pervenire, quod Avicenna quoquo dixit in libris meteorlogicis, lumen vocans vehiculum virtutum omnium caelestium et Albertus in libro *de somno vigiliaque* confirmavit.' To finish the translation of the quote begun above, in English: 'which Avicenna also stated in his book on meteorology, calling light the vehicle of all power of the heavens, and Albert confirmed [this] in his book about sleep and wakefulness.'

[104] Wayne Shumaker, *Occult Sciences in the Renaissance: A Study in Intellectual Patterns* (Berkeley: University of California Press, 1979), p. 19.

[105] The notion of affinities and antipathies was based on belief in a completely interconnected universe. Celestial bodies affected or repelled earthly objects, depending upon whether the object in question was indirectly, meaning 'sympathetically,' attached or 'antipathetically' opposed to the celestial body. For Pico's warning that such beliefs could lead one into superstition, see Shumaker, pp. 22–23.

[106] Shumaker, pp. 22.

[107] Rabin presents a counterargument. See, Sheila Rabin, 'Unholy Astrology: Did Pico Always View it That Way?', in *Paracelsian Moments' Science, Medicine & Astrology in Early Modern Europe*, ed. Gerhild Scholz Williams and Charles D. Gunoe, Jr. (Kirksville, MO: Truman State University, 2002): pp. 151–162, pp. 158–162. She bases her argument on statements made by Pico such as this one: 'since an astrologer looks at signs that are not signs, and thinks about causes that are not causes, he is, therefore, deceived' (quoted on p. 158). However, the

to be concerned that astrologers will lead people to focus upon worldly forces rather than maintaining an attentive regard for God.[108] This is much the same concern that had led Gerson earlier to reject astrology; the fear that it might lead those who practiced and put faith in it into idolatry.

In sum, Pico did not reject the central premise upon which astrology was built, the idea that celestial influences affect terrestrial creatures and events, nor did he offer a clear rejection of the possibility of astrological divination. To have done so would have been to offer a serious challenge to the cosmological system infusing the worldview of educated Europeans of his day. Furthermore, one will look in vain to find a clear denunciation of astrology and its principles in the sixteenth or even early seventeenth century. In fact, even Johannes Kepler (1571–1630) – a contemporary and occasional correspondent of Galileo – wrote a defense of astrology in the seventeenth century. In terms very reminiscent of Albert the Great, Kepler stated flatly that the stars incline, but do not compel, people toward certain behaviours or personality characteristics.[109] To provide a descriptive example of how such inclination works, he argued that Saturn in the ascendant of Count Albrecht von Wallenstein's (1583-1634) natal chart, in conjunction with the Moon's position, accounted for Wallenstein's terrible character.[110] As for those who might question the efficacy of astrological forecasting, he pointed to what he considered empirical proof of the accuracy of astrological predictions. For example, in his *Third Man in the Middle* he argued that his past weather predictions demonstrated that a conjunction between Mars and the Sun resulted in unseasonably warm weather, pointing to the winters of 1598 and 1601 where this condition

---

evidence suggests that Pico rejected certain astrological practices, but not astrology as a whole.

[108] For a much more comprehensive consideration of Pico's attitude toward astrology see Rutkin, 'Astrology, Natural Philosophy and the History of Science' OR IS IT 'Galileo Astrologer', pp. 230–305.

[109] Sheila J. Rabin, 'Kepler's Attitude Toward Pico and the Anti-Astrology Polemic,' in *Renaissance Quarterly* 50, no. 3 (1997): pp. 750–770, p. 753. For more on Kepler's astrological beliefs, see J. Bruce Brackenridge and Mary Ann Rossi, 'Johannes Kepler's "On the More Certain Fundamentals of Astrology" Prague 1601', *Proceedings of the American Philosophical Society* 123, no. 2 (1979): pp. 85–116; Gerard Simon, *Kepler: Astronome, Astrologue* (Paris: Gallimard, 1979); Dorian Gieseler Greenbaum, ed., 'Kepler's Astrology', *Culture and Cosmos* 12, nos. 1 and 2 (Spring/Summer and Autumn/Winter 2008).

[110] Rabin, 'Kepler's Attitude Toward Pico and the Anti-Astrology Polemic', p. 752.

caused mild winters and thunderstorms.[111] True, the association between weather and celestial influences is indicative of a belief in natural astrology, but Kepler also referred to celestial influence over human nature – Wallenstein – suggesting a broad acceptance of astrological principles in all their variety.

Even though Kepler rejected the central geocentric tenet of the Aristotelian-Ptolemaic cosmos, he did not reject the unifying notion of celestial influence that had held a central place in the worldview of European intellectuals for centuries. Instead, he still accepted a cosmological model in which all things are interconnected and the human body was presumed to be a microcosmic representation of the larger macrocosm of the universe. Celestial motion, combined with the qualities of individual celestial objects, affected the four humours of the human body through emitted rays. We should keep in mind why and in what ways this was a rational model for understanding the world: astrology represented a model of the universe in which physical bodies interacted with one another with perfect, mathematically describable consistency, requiring no belief in invisible entities or divine intervention to explain everything from the movements of the planets to the coming to be and passing away of terrestrial objects. Understanding this point is important, for it represents a perfect example of the way in which basic beliefs functioning conjunctively with beliefs logically developed or derived from them can lead to a thoroughly rational system operating within different parameters than those established by modern science.[112]

This point has been regularly misunderstood, in part again due to misinterpretations of the historical record that seem to be driven by the distaste for astrology evidenced by some writers.[113] The resultant misunderstandings have blurred our perception of the past, nowhere more so than when we consider an iconic figure such as Galileo. Whereas the 'Galileo legend' presents a man who is a persistent opponent of astrology – or who rejected astrology late in life – the truth was far different. As we saw earlier, he cast horoscopes in the 1580s and 90s. Even in 1610 we find

---

[111] Rabin, 'Kepler's Attitude Toward Pico and the Anti-Astrology Polemic', p. 751.

[112] Richard Swinburne addresses both 'basic beliefs' and the 'conjunctive nature of ideas' in his *Faith and Reason*, pp. 3–24.

[113] For an example, see Paul Oskar Kristeller's treatment of Pico della Mirandola in *Eight Philosophers of the Italian Renaissance* (Stanford: Stanford University Press, 1964), p. 68.

this affirmation of the influences of heavenly bodies in the opening of the *Siderius nuncius:*

> So who does not know that clemency, kindness of heart, gentleness of manners, splendor of royal blood, nobleness in public functions, wide extent of influence and power over others, all of which have fixed their common abode and seat in your highness - who, I say, does not know that these qualities, according to the providence of God, from whom all good things do come, emanate from the benign star of Jupiter?[114]

One could choose to read such a statement as a piece of standard rhetoric, an instance of Galileo telling his correspondent what he expected to hear. But to make such an assumption would be to remove Galileo from his historical context. For one thing, this comment was no simple one-off statement. Instead, as Mario Biagioli has demonstrated, 'astrological predetermination' was a repetitive refrain in Galileo's communications with the Medici, as he sought to demonstrate how his own findings confirmed the Medici's destiny.[115] For another, Galileo did not merely make use of flattering rhetoric – he also made regular and sustained use of astrological terminology in ways that indicate he was knowledgeable about the subject. For example, turning back to *The Starry Messenger*, Galileo wrote this encomium to the way in which the Medicean stars demonstrated the inevitability of the Medicean rise to power in Florence:

> Jupiter, Jupiter, I say, at the first rising of your highness, had already passed beyond the torpid vapours of the horizon, and, occupying the middle of the sky and illuminating with his royal seat the eastern corner, looked over your blessed birth from his sublime throne; he poured out all his splendour and grandeur into the most pure air, so that the tender body and the soul (decorated by God with the most noble ornaments) should drink, with the first breath, his universal strength and power.[116]

This was not a man who was merely employing hyperbole in pursuit of patronage. This was a man who thoroughly understood astrology's tenets and employed them in his writings throughout his life. Rather than being a

---

[114] Galileo Galilei, *Sidereus Nuncius, or The Sidereal Messenger*, trans. Albert Helden (Chicago: University of Chicago Press, 1989), p. 31.

[115] Mario Biagioli, 'An Astrologico-Dynastic Encounter', *Galileo's Astrology*, special issue of *Culture and Cosmos* 7, no. 1 (2003): pp. 59–63, p. 60.

[116] Galilei, *Sidereus Nuncius,* in *Opere*, III.1, p. 56, though I am using Ernst's translation, from p. 28.

sceptic of the discipline, he was, if anything, overly confident in the abilities of judicial astrology to predict the future, assuming that he was guilty of even half the level of astral-fatalism the inquisition accused him of in 1604.[117]

And why should Galileo not have felt confident in judicial astrology's predictive powers? As we saw above, centuries of debates and natural philosophical assumptions had gone into creating a worldview in which the notion that celestial influences affect terrestrial objects was a central component one that was widely, and perhaps universally, accepted. Therefore, the conceptual framework 'forced upon' Galileo by his experiences of the world, as Richard Swinburne puts it,[118] was one in which heavenly bodies *must* impart influences to terrestrial creatures. Given that view, all that was required for a proper prediction was a thoughtful evaluation of the evidence coupled with the right level of mathematical skill – and no one has ever been more confident in his abilities than Galileo. Within his worldview, it would have been irrational for him not to have accepted the validity of judicial astrology.

Galileo even strove to put his discoveries within the context of this worldview. On 21 May 1611 Galileo responded to a letter from Archbishop Piero Dini in Rome.[119] Dini asked Galileo how his 'Medicean stars', the four moons of Jupiter Galileo had discovered, might influence humans. Galileo's response takes up eleven pages in Favaro's edition of Galileo's works. Stating that it would be wrong to imagine that 'these Medicean Planets lack all influence, wherein the other stars abound'[120] in such influences, Galileo then goes on to speculate what that influence might be:

---

[117] Antonino Poppi, 'On Trial for Astral Fatalism: Galileo Faces the Inquisition,' *Galileo's Astrology*, special issue of *Culture and Cosmos*, 7.1 (2003): pp. 49–58.

[118] Swinburne, *Faith and Reason*, p. 17.

[119] Mark Edwards, 'Galileo's Letter to Piero Dini, Rome 21 May 1611', *Culture and Cosmos* 7, no. 1 (2003): pp. 85–95; Galileo Galilei, 'letter of 21 May, 1611', in *Edizione nazionale delle opere di Galileo Galilei*, ed. A. Favaro (Florence: G. Barbèra, 1888), Vol. 11, pp. 105–116; Nick Kollerstrom, 'Galileo's Astrology', *Largo campo di filosofare: Eurosymposium Galileo 2001*, ed. José Montesinos and Carlos Solís Santos (La Orotava: *Fundacion Canaria* Orotava de Historia de la Ciencia, 2001): pp. 421–432. I am relying on Mark Edwards' translations here, checked against the original. I am thankful for the assistance of Doctor Massimo Rondolino in this matter.

[120] Galilei, 'letter to Piero Dini, p. 105; Kollerstrom, 'Galileo's Astrology', p. 425.

If, therefore, of the inferior causes, those which arouse boldness of heart are diametrically contrary to those which inspire intellectual speculation, it is also most reasonable that the superior causes (if indeed they operate on us) be utterly different from those on which courage and the speculative faculty depend; and if the stars do operate and influence principally by their light, perchance it might be possible with some probable conjecture to deduce courage and boldness of heart from very large and vehement stars, and acuteness and perspicacity of wit from the thinnest and almost invisible lights.[121]

These are certainly not the words of a man who questions the commonly accepted notion of his day, that heavenly bodies influence terrestrial ones. Again, we could speculate along with Mario Biagoli that Galileo might be making such statements for this or that reason of policy or rhetoric – but where is the evidence?[122] The simplest answer, demanding the fewest assumptions, is that Galileo – the man who had cast dozens of horoscopes for friends, family, and patrons, the man who regularly employed astrological motifs in his writing – was a man who truly believed in the divinatory power of astrology.

Furthermore, it does not appear that Galileo's astrological interests were in any way peripheral or in opposition to his work as a scientist. Bernadette Brady has provided a thoughtful and complex analysis of 'Galileo's Astrological Philosophy', and she has argued that his study of astrology may well have provided the impetus for his break with the Aristotelian cosmological model that portrayed the heavens as perfect and unchanging in contrast to the imperfect and changeable sublunar sphere. She argues that the 'pragmatic, predictive-seeking, axiomatic approach to the physical world'[123] provided by the astrologers of the Arabic speaking world allowed Galileo to see the supra and sublunar spheres as linked and obedient to the same laws of nature.[124] In this analysis, the Arabicized astrology that influenced Galileo's own astrological practices acted as a spur rather than an impediment to his developing scientific worldview.

Keeping all of the above information in mind, it is important to recognize that the context of seventeenth-century Europe would make any

---

[121] Galilei, 'letter to Piero Dini', p. 111; Kollerstrom, 'Galileo's Astrology', p. 425.
[122] Biagioli, *Galileo, Courtier*. Biagioli refers to Galileo's 'astrological rhetoric' frequently, such as on p. 115 and p. 128.
[123] Brady, 'Galileo's Astrological Philosophy', p. 80.
[124] Brady, 'Galileo's Astrological Philosophy', p. 97.

acceptance of predictive astrology and its principles fully rational. In fact, it would be well after Galileo's death in 1642 that astrology would begin to fall out of favour with the European intellectual elite due to associations with riot, rebellion, and the uneducated that developed during the English Civil Wars (1642–1648), an attitude that then spread to the continent.[125] Even then, as late as 1799 it was required that the chair of *astronomia* at the University of Bologna produce an annual almanac for the use of the medical students and faculty.[126] Therefore, if we are to understand Galileo and the work he did within his historical context, then we should also understand that the conceptual framework within which he worked was very different from that of a modern scientist. His was one in which concepts of celestial influences and the effects thereof had been debated and discussed for centuries. The intellectual consensus of Galileo's day was so firmly in favour of the effectiveness of astrology that its principles held hegemonic status and it would not have occurred to him to question the idea that the heavens influence terrestrial events. Given that starting point, determining the future influences of the heavens was simply a matter of mathematical astronomy.

Although those who study the history of astrology are already well aware of Galileo's astrological work, it is well past time for the broader body of scholars studying Galileo to recognize these facts. Even though Galileo rejected the geocentric cosmological view in favour of heliocentrism, there was no reason to scrap the centuries-old notion that celestial bodies influenced terrestrial events or that one could then make predictions about the future based on a study of the heavens. His near contemporary Kepler certainly did not. In fact, astrological belief may have fruitfully influenced astrologers such as Kepler and Galileo, as Brady has argued regarding Galileo. After all, the essence of these beliefs was that heavenly bodies moved and interacted with one another in mathematically describable and ordered fashion, in ways 'which could be calculated, measured and tabulated'.[127] Additionally, astrologers had long been

---

[125] Hendrix, *How Albert the Great's* Speculum Astronomiae *was Interpreted and Read*, pp. 205–230; For the English civil wars see Patrick Curry, *Prophecy and Power: Astrology in Early Modern England* (Oxford: Polity Press, 1989).

[126] Tester, *A History of Western Astrology*, p. 184. It is true that the late sixteenth century saw the emergency of similar associations between violence and unrest and astrology in France, but these ideas did not appear to become influential in other regions until the following century. See Denis Crouzet, *Les Guerriers de Dieu* (Seyssel: Champ Vallon, 1990), Vol. I, pp. 101–304.

[127] Brady, 'Galileo's Astrological Philosophy', p. 97.

proponents of careful observation and the gathering of data.[128] Therefore, with Galileo working within the context of a newly emerging mechanical philosophy dependent on empirical observation, there is no reason to imagine that astrological beliefs and practices would have acted as an impediment to his work – and by reinforcing the importance of empirical observation coupled with mathematical precision, astrology may have helped to promote those ideas.

Acknowledgements

I would like to thank the reviewers who read over the first version of this article, for their many thoughtful and helpful comments and suggestions.

---

[128] Many practicing astrologers relied on charts and tables compiled by others, as Pierre Gassendi (1592–1655) dismissively stated. But even then, such tables and charts were repositories of data earlier astrologers had compiled. See Tester, *A History of Western Astrology*, pp. 230–232. Furthermore, there were plenty of early modern astrologers who made their own observations. Tycho Brahe (1546–1601) provides just such an example. Victor E. Thoren, *The Lord of Uraniborg: A Biography of Tycho Brahe* (Cambridge: Cambridge University Press, 1991).

Paradise Lost and the Descent of Urania: from Astrology to Allegory

Richard Angelo Bergen

Abstract: John Milton's cosmic portraiture in *Paradise Lost* is relevant to the nature, function and meaning of the actors therein, and this essay aims to analyse one such inhabitant: Urania, Milton's muse. Milton calls upon Urania to govern his song, and he does so in the context of a cosmologically charged set of epic books, raising her traditional associations as the goddess of astronomy, and linking her with music. However, he also radically Christianizes/allegorizes Urania, ousts her from her home in the crystalline spheres, and exorcises her of astrological powers. Accordingly, this essay describes Ptolemaic cosmology and Urania's place within it, treating Bernardus Silvestris's *Cosmographia* as representative in this respect. My essay argues that Milton rejects the old view, using the *Cosmographia* as a counterpoint. I draw on a new crop of Milton scholarship attentive to the influence of the new astronomy on the epic poet, adding my own adumbrations to their findings. Milton's invoking 'the meaning not the name,' his interest in the semiotics rather than the substance of Urania, is concomitant with rejecting the cosmos in which she was said to have her place as a celestial intelligence.

Many Miltonists of eminent scholarly stature, over the past century and more, have remarked on Urania, so it is not without forwardness that I presume to say something *new* about John Milton's muse. I hope to do this, dialectically, by reflecting on the *old* traditions of Ptolemaic astrology associated with her, and by noting that the current scholarship on Milton and the new cosmology (which elides explicit discussions of Urania for the most part) has tremendous implications for understanding the manner in which Milton is appropriating Urania. Since 2012, Karen Edwards, Malabika Sarkar, John Leonard, and Dennis Danielson have authored magisterial interventions into the subject of Renaissance Cosmology and *Paradise Lost*, and I intend to point out the pertinence of their viewpoint to the scholarly discussion of Urania, with my own adumbrations as well.[1] To

---

[1] Dennis Richard Danielson, *Paradise Lost and the Cosmological Revolution* (Cambridge: Cambridge University Press, 2014); Karen Edwards, 'Cosmology', in

---

Richard Angelo Bergen, '*Paradise Lost* and the Descent of Urania: from Astrology to Allegory', *Culture and Cosmos,* Vol. 18, no. 2, Spring/Summer 2014, pp. 105-124.
www.CultureAndCosmos.org

this end, I detail Urania's 'descent' in the earlier cosmology at some length, and discuss Bernardus Silvestris' *Cosmographia* as a poem presenting the apogee of Urania's stature as an astrological divinity, offering it as a counterpoint to *Paradise Lost*. A key principle is that Milton's cosmic portraiture in *Paradise Lost* applies saliently to the nature, function and meaning of the actors of its cosmos. Milton calls upon Urania to govern his song, and he does so in the context of a cosmologically charged set of epic books, raising her traditional associations as the goddess of astronomy. He links her with music, and refers to her as a goddess of wisdom in the heights of heaven, all appropriate to her role in the old cosmological picture. However, he also radically Christianizes/allegorizes Urania, ousts her from her home in the crystalline spheres, and exorcises her of astrological powers.

From whence does Urania descend? The beginning of Urania and her sisters, the nine classical muses, is recorded briefly by Hesiod in the *Theogony*: they are the daughters of Mnemosyne (Memory) and Zeus, set perpetually upon song, and dwelling near the top of Olympus.[2] H. David Brumble notes that in other traditions the muses were the daughters of Harmonia, further associating them with divine music.[3] Homer's invocations to other muses in his epics also served to entangle the collective with the enterprise of epic poetry. Urania's name can be translated from the Greek as 'sky' or 'the heavens,' and concomitantly, in Plato's *Symposium*, Pausanius relates her to proportion, astronomy and astrology; in fact, already in this Platonic text, Urania transcends her role as one of many singing muses, to become the 'motherless,' heavenly aspect of Aphrodite, representing divine intellect and the 'heavenly species

---

The Cambridge Companion to Paradise Lost (Cambridge: Cambridge University Press, 2014), pp. 109–24; John Leonard, *Faithful Labourers: a Reception History of* Paradise Lost*, 1667-1970* (Oxford: Oxford University Press, 2013), pp. 705–819; and Malabika Sarkar, *Cosmos and Character in Paradise Lost* (New York: Palgrave Macmillan, 2012). Arguably, this new current in criticism began earlier with, for instance, Catherine Gimelli Martin, "'What if the Sun Be Centre to the World?': Milton's Epistemology, Cosmology, and Paradise of Fools Reconsidered', *Modern Philology* 99, no. 2 (2001): pp. 231–65.
[2] Hesiod, *Theogony* and *Works and Days*, trans. M. L. West (Oxford: Oxford University Press, 1988), p. 5.
[3] H. David Brumble, *Classical Myths and Legends in the Middle Ages and Renaissance: A Dictionary of Allegorical Meanings* (Westport, CT: Greenwood, 2013), p. 228.

of love'.[4] Moreover, in Book X of the *Republic*, Plato writes of eight sirens that accounted for the *musica universalis* (emerging from a sense of the harmony in the structure of the heavens), to be later identified with the nine Muses. Milton himself pictures the nine sirens that 'sit upon the nine enfolded spheres' and 'turn the adamantine spindle round, / On which the fate of gods and men is wound' in his early masque, *Arcades* (ll.64, 66-7).[5] The muses were in this way lifted from the Greek mountaintop into the higher spheres of a philosophical cosmology.

In speaking about who Urania 'is' we have already begun to discuss her place in the old world-picture of the 'spheres.' After Plato, Aristotle theorized that the universe was based upon the principle of sphericity; in *De Caelo*, Aristotle wrote that the cosmos was based on primary shapes in nature, with the four elements (earth, water, air, fire) exhibiting rectilinear motion and heavenly bodies appearing circular in their movements. While the earth was constituted by the four elements (especially earth and water, the heaviest of the four elements in 'nature'), it was nevertheless the first sphere in the upward climb through the cosmos. The outer parts of the globe were followed by a sphere of air, then fire. The flux of these elements caused the mutability experienced by mortals on the terrestrial orb. The successive concentric spheres moving out from earth were all 'super-lunary' – above the moon, above nature, moving in an environment of quintessence, or *aether* – and were unchanging, ostensibly evidenced by the planet's perpetual orbit on a crystalline sphere. Aristotle was deeply impressed that it was proper to planets that they should move circularly, for circles were 'prior in nature to, and therefore more divine than, the four known elements':

> Of the simple lines, only the circle is perfect or complete. (A straight line must be either infinite or finite. If infinite it can by definition never be complete, if finite it is always capable of extension and therefore cannot be said to be complete.) But the perfect is prior in nature to the imperfect. Hence circular motion is prior to rectilinear, and the body whose natural motion it is must therefore be prior in nature to the bodies which move in straight lines.[6]

---

[4] Plato, *Symposium*, in *Complete Works*, trans. John M. Cooper, D.S Hutchinson, and Jonathan Barnes (Indianapolis, IN: Hackett, 1997), 180d-187e.

[5] All quotations of Milton are from John Milton, *Complete Poems and Major Prose*, ed. Merritt Y. Hughes (New York: Odyssey, 1957).

[6] Aristotle, *On the Heavens*, trans. W. K. C. Guthrie (Cambridge, MA: Harvard University Press, 1939), I.ii.

If terrestrial life was subject to decay because of the contrary motions of the four elements, the planets were incorruptible and eternal because they encountered no kinetic resistance in a realm above the nature of these corruptible elements, in the motion proper to their super-nature: 'Thus the reasoning from all our premises goes to make us believe that there is some other body separate from those around us here, and of a higher nature in proportion as it is removed from the sublunary world.'[7] Here, in the realm of super-nature, the aether, the muses reside.

The order of the spheres outside the earth was: the sphere of the Moon, Mercury, Venus, the Sun, Mars, Jupiter, Saturn, the sphere of the fixed stars, and the *primum mobile* (the ninth sphere of the heavens, a proper crystalline sphere, was posited in the fourteenth century to account for what went on in 'leap years').[8] The nine muses were to occupy their respective spheres of the heavens, up to the fixed stars (Thalia was usually consigned to earth). Martianus Capella, for instance, wrote that

> [t]he upper spheres and the seven planetary spheres produced a symphony of the harmonious notes of each... indeed... [each Muse] took her position where she recognized the pitch that was familiar to her. For Urania was attuned to the outermost sphere of the starry universe, which was swept along with a high pitch.[9]

Although Calliope was sometimes considered the preeminent among the muses, Urania often occupied the highest place. The muses account for the harmonious symphony of the universe, but Urania is the tuner and conductor, 'governing of all the celestial motions and their harmony,' as the prominent Italian humanist Patrizi wrote in *L'amorosa filosofia*.[10] Guillaume du Bartas, whom we will turn to again, also states in *L'Uranie* that Urania's '*Nine*-fold Voice did choicely imitate / Th' *Harmonious*

---

[7] Aristotle, *On the Heavens*, I.iii.

[8] Chauncey Wood, *Chaucer and the Country of the Stars: Poetic Uses of Astrological Imagery* (Princeton, NJ: Princeton University Press, 1970), pp. 58–60.

[9] Martianus Capella, *Martianus Capella and the Seven Liberal Arts*, trans. Richard Johnson and William Harris Stahl (New York, NY: Columbia University Press, 1977), p. 16.

[10] Translated in Jacomien Prins, *Echoes of an Invisible World: Marsilio Ficino and Francesco Patrizi on Cosmic Order and Music Theory* (Leiden: Brill, 2015), p. 379.

Musik of *Heauens* nimble Dance' (Stanza 9).[11] This is the dignity of Urania, the astrological divinity.

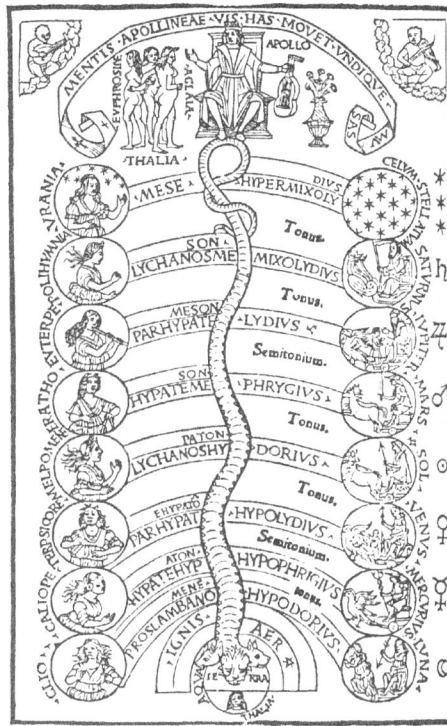

Fig. 1. Gafurius's *Practica musice*, 1496. The muses are assigned to each their respective planetary spheres. The muses were said to generate their music of the universe through the impetus of the *primum mobile*, and on the right column, the increments of the musical scale correspond to the space between spheres.

One is perhaps tempted to assume that the writer is speaking figuratively. However, the medieval worldview understood a *substantive* analogy between the cosmological reality and the anthropological reality, or a relationship between microcosm and macrocosm. The early Renaissance Neo-Platonist Heinrich Cornelius Agrippa was not being one-sidedly metaphorical when he stated that

---

[11] Guillaume de Saluste Du Bartas, *The Divine Weeks and Works of Guillaume Du Bartas*, trans. Joshua Sylvester, ed. Susan Snyder (Oxford: Oxford University Press, 1979).

The measures of all the members [of man's body] are proportionate, and consonant both to the parts of the world, and measures of the Archetype, and so agreeing, that there is no member in man which hath not correspondence with some other sign, Star, intelligence, divine name, sometimes in God himself the Archetype.[12]

In this same work, Agrippa explicitly connects Urania to the divinely instituted intelligence dwelling in the fixed stars.[13] The muses, in fact, were translated from Olympian deities, to sirens, to heavenly-sphere-dwelling musicians, and underwent another symbolic translation to 'intelligences,' but they were never *merely* symbolic. Many intellectuals did believe in the *intelligences* of the spheres as instituted by God, a view which appears to have held some respectability by the time of Bernardus Silvestris (ca. 1085–1178 CE).[14] Plato expresses a belief in souls of planets in his massively influential *Timaeus*: the Supreme being, Plato writes,

divided [the soul of the universe] into a number of souls equal to the number of the stars and assigned each soul to a star. He mounted each soul in a carriage, as it were, and showed it the nature of the universe. He described to them the laws that had been foreordained.[15]

---

[12] Heinrich Cornelius Agrippa Von Nettesheim, *Three Books of Occult Philosophy*, trans. John French (London: Printed by R.W. for Gregory Moule, 1651), p. 264. Although Agrippa was a unique and somewhat heterodox intellectual (not representative of Renaissance opinion in all respects), in this matter he is presenting a fairly conventional cosmological view.

[13] Agrippa, *Three Books*, p. 333

[14] For background on medieval cosmology and intelligences, see Nicholas Campion, *A History of Western Astrology Volume II: The Medieval and Modern Worlds* (Cornwall: Continuum, 2009), pp. 29–58; and C. S. Lewis, *The Discarded Image: an Introduction to Medieval and Renaissance Literature* (Cambridge: Cambridge University Press, 1964), pp. 91–122. A scholar providing a more specific discussion of the muses as intelligences: Stella P. Revard, "'L'allegro' and 'Il Penseroso': Classical Tradition and Renaissance Mythography', *Publications of the modern Language Association of America* 101, no. 3 (1986): pp. 338–50. For more discussions on intelligences, see, for instance, Nicolas Weill-Parrot, 'Astrology, Astral Influences, and Occult Properties in the Thirteenth and Fourteenth Centuries' *Traditio* 65 (2010): pp. 201–230; Edward Grant discusses intelligences at several points in *Planets, Stars, and Orbs: The Medieval Cosmos, 1200-1687* (Cambridge: Cambridge University Press, 1996), although he downplays the 'animism' of what medieval thinkers held to be celestial intelligences, particularly after 1277 CE.

[15] Plato, *Complete Works*, 41d-e.

The 'souls' Plato mentions were held responsible for the arbitration of intelligence, and the reality that material things were given form: 'intelligence depends on the intelligence of the heavens,' writes Ficino, and,

> All the intelligences, be they those of the highest rank and superior to the souls, or be they inferior and part of the souls, are so interconnected that, beginning with God who is their head, they proceed in a long and uninterrupted chain, and all the superior ones shed their rays down on the inferior ones.[16]

Milton's contemporary George Sandys also acknowledged this cosmic position for the muses, although with slightly more distancing language. 'The Muses are also taken for the Intelligences, of the Coelestiall Spheares; which... doe make a diversity of sounds; and consequently (according to Pythagoras) an incredible harmony. Yet this saith Macrobius is not to be heard, in that so vast a sound cannot enter at the narrow labyrinth of the eare.'[17] The traditional arrangement of the muses was still a commonplace in the mid-seventeenth, if the ontological status of the sisters had then become more a matter open to question.

As the highest celestial intelligence affecting the operations of the cosmos, Urania was a divinity – an *astrological* divinity – in the later adaptations of the Ptolemaic model, dwelling in the sphere of the fixed stars. Although astrology was focused on the seven planets as they figured on the zodiac, the whole system of belief represented by the Ptolemaic system and its spheres accommodated a space for the muses, and for Urania, above the planets. Bernardus Silvestris's *Cosmographia*, a Neo-Platonic as well as Ptolemaic mid-twelfth-century narrative of creation, is an excellent imaginative expression of the old cosmos, and will serve as an apt counterpoint to Milton's epic. The first portion of the poem involves Natura complaining to Noys (Divine Providence) that Hyle (primordial matter) is prodigal and disorderly; and so, Noys separates the elements of the universe and proceeds to create creatures and stars according to their kinds; this takes place in the first half of the poem, entitled Megacosmos. The second half of the poem, Microcosmos, pertains to the creation of

---

[16] Eugenio Garin, *Astrology in the Renaissance: The Zodiac of Life*, trans. Carolyn Jackson and June Allen (London: Routledge & Kegan Paul, 1983), p. 70.
[17] George Sandys, *Ovid's 'Metamorphosis' Englished, Mythologized, and Represented in Figures* (Oxford: John Lichfield, 1632), p. 248.

mankind, in which Noys commissions Natura to seek out Urania, who is above the seven planetary spheres. Noys states that Nature is incapable of the 'generation of the human soul, and the creation or instillation in this soul of the radiance of eternal vitality,' and so must search out Urania who is 'in close attendance upon my throne'.[18] Urania is 'concerned with the whole sky and all the stars',[19] and thus Natura has some difficulty finding her in the vast abode of the fixed stars. At last, after travelling outside the seven planetary spheres, and circling through much of the zodiac, she discovers Urania in dazzling brightness, 'gazing in wonder at the heavens, calculating their recurrent motions and periods of their orbits, in accordance with sure standards of observation',[20] clearly corroborating Urania's patronage of astrology. Urania agrees to descend with Natura to earth in order to confer intellect, memory, and knowledge of the heavens upon man: thus the two descend through each of the planetary spheres, the sphere of fire and of air, encountering all of the planetary deities and discoursing on their attributes. Significant, given the parallels to *Paradise Lost*, Natura and Urania observe a 'Golden Chain' along the lunar boundary descend to earth, 'which unties the higher with the lower universe.'[21] The two descend to earth, and with the help of Physis as well as long discussion and process, they create mankind: Urania crafting the intellective and spiritual properties of man and expelling 'the evil taint from Silva' and containing 'fluid matter within fixed bounds'.[22] For Silvestris, then, Urania the astrological divinity occupies an indispensable role in his narrative, and simultaneously fills an important role in this cosmos.

Although Natura and Urania's journey conforms to some epic motifs, Silvestris does not explicitly invoke any muse – Urania's older status as a muse has faded into the background to make way for her still-musical, effulgent and three-dimensional divinity. Milton restores Urania's old function as a muse for poetry in *Paradise Lost*; however, the differences between Milton's Urania and Silvestris' are enormous, and we will proceed to consider Milton's radical Christianization/allegorization of Urania, as well as Milton's different cosmology and the implications it has for Urania's nature and function. At first glance, Milton's usage of Urania

---

[18] Bernardus Silvestris, *The Cosmographia of Bernardus Silvestris*, trans. Winthrop Wetherbee (New York: Columbia University Press, 1973), p. 94.

[19] Silvestris, *Cosmographia*, p. 95.

[20] Silvestris, *Cosmographia*, p. 97.

[21] Silvestris, *Cosmographia*, p. 105.

[22] Silvestris, *Cosmographia*, p. 118.

is, as E. R. Gregory argues, 'conventional and unoriginal'.[23] Milton needs a figure that his readers will recognize and respond to; all epic poets need a muse, and Milton needs one from the 'Heav'n of Heav'ns' (7.13), one that has less to do with 'long and tedious havoc fabl'd Knights / In Battels feign'd' (9.30-31). Happily, 'Urania's essence, whether portrayed in an averted posture or with her eyes turned toward the heavens, was renunciation, a frank indifference to the things of this world'.[24] Mindelle Anne Treip outlines how Renaissance visual artists associated Urania iconographically with astronomy, poetry, theology, philosophy and contemplation,[25] all of which suit the themes of Milton's poem extraordinarily well.[26] In Milton's original 10-book design, the invocation to Urania would have been in the same book as Raphael's discourse on astronomy, resonating with Urania's role as the patroness of that subject. Moreover, the invocation in Book VII is saturated in language pertaining to sonority, appropriate for a muse conducting the symphony of the stars.

Perhaps most significantly, European Christians had been using 'Urania' as a signifier of sacred poetry for some time. Edmund Spenser's *The Teares of the Muses* refers to Urania as '[t]he heavenlie light of knowledge'.[27] Drummond of Hawthornden's collection of devotional material was entitled *Urania, or Spiritual Poems*. According to John Shawcross, it was sometimes conventional for Renaissance humanists to interchange the Holy Spirit with Urania, and supports his point by citing Samuel Austin's prayer, 'Thy Spirit bee my Urania', as in his book, *Urania, or, The Heavenly Muse*.[28] Many scholars point to Guillaume Du Bartas' poem *L'Uranie*, which Milton probably read in the immensely

---

[23] E. R. Gregory, *Milton and the Muses* (Tuscaloosa, AL: University of Alabama, 1989), p. 69.

[24] Gregory, *Milton and the Muses,* p. 114. We have already noted Urania's heavenward posture in the *Cosmographia*.

[25] Mindele Anne Treip, *Descend from Heav'n Urania: Milton's Paradise Lost and Raphael's Cycle in the Stanza Della Segnatura* (Victoria: University of Victoria, 1985), p. 24.

[26] Darcy Woodall identifies typical paraphernalia to include a starry veil, a crown with constellations, a sphere or globe, and a staff or compass. 'Urania: Transcendent Muse', *Spring: A Journal of Archetype and Culture* 70 (2004): p. 167.

[27] Edmund Spenser, *The Complete Works in Verse and Prose of Edmund Spenser, Vol. 3*, ed. Alexander Grosart (London: Spenser Society, 1992), 1.488.

[28] John Shawcross, 'Urania,' in *A Milton Encyclopedia*, ed. William B. Hunter (Lewisburg, PA: Bucknell University Press, 1978), 112.

popular translation of Joshua Sylvester; Milton borrowed a line from *Divine Weeks and Works* (Sylvester's translation containing Du Bartas' creation poem and *L'Uranie*): *Paradise Lost* 3.373, 'Immutable, immortal, infinite'.[29] Du Bartas' poem was published in an preponderantly religious collection of poetry entitled *La Muse Chrestienne* (1574), and the Urania persona of *L'Uranie* urges the poet to compose verse on Scriptural themes. John Mulryan demonstrates at considerable length that almost all classical-mythical appropriation in the Renaissance involved allegory,[30] and we see in these examples a thoroughly Christian pattern of figuration for Urania – which, however, is not quite to say that these authors rejected the idea of Urania as an astrological divinity. Urania, the speaker in part of Spenser's poem notes the influence of the stars on the minds of humans, and exults in the cosmological features of the 'Christall firmament,' 'heaven's great Hierarchie,' and 'The Spirites and Intelligences fayre,' even while she herself appears mainly to represent a certain kind of attitude and principle, utterly deferent to God.[31] Indeed, Silvestris's and Patrizi's Urania is a quasi-Christian figure as well, even if she functions less *figuratively* than in Du Bartas.

Indeed, the scheme of the cosmos that Du Bartas paints is quite close to Silvestris, and Du Bartas certainly leaves no doubt about fortuitous astrological belief:

> Sens-less is he [...]
> that doth affirm the Stars
> To have no force on these inferiours;
> Though heav'ns effects we most apparant see
> In number more then heav'nly Torches be.[32]

---

[29] Du Bartas, *The Divine Weeks and Works of Guillaume Du Bartas*, 1.1.56.

[30] John Mulryan, *Through a Glass Darkly: Milton's Reinvention of the Mythological Tradition* (Pittsburgh, PA: Duquesne University Press, 1996). Mulryan points out the legions of scriptural and mythological commentators, classical dictionaries, emblem books and *Ovid Moralisé* books that were almost entirely comprised of allegorical explanations, and links Milton to them. Representative texts I would point to include Sandys' book on Ovid, cited earlier, and Francis Bacon's *De Sapientia Veterum.* Another work connecting Renaissance thought on the classics to allegory, and then to Milton, is Kenneth Borris, *Allegory and Epic in English Renaissance Literature: Heroic Form in Sidney, Spenser, and Milton* (Cambridge: Cambridge University Press, 2000), chapters 1, 2, 3, 6 and 7.

[31] Edmund Spenser, *The Tears of the Muses*, ll.506-509.

[32] Du Bartas, *The Divine Weeks and Works of Guillaume Du Bartas*, 1.4.456-63.

Still, Du Bartas' Urania feels quite different from Silvestris': her primary function is to exhort the writer to poetic piety, eloquence and biblical fidelity, not to urge astronomical study, or to act as a celestial intelligence or subsidiary agent of creation. While Silvestris sends his Urania straight to the Empyrean, Du Bartas, in discussing the Empyrean timidly backs down: 'Nor shall my faint and humble Muse presume / So high a Song and Subiect to assume'.[33] This passage is problematic given that elsewhere Du Bartas' Urania very nearly identifies herself with the Holy Ghost:

> Let CHRIST (as *Man-God*) be your *double Mount*
> Whereonto *Muse,* and, for the winged hoove
> Of *Pegasus,* to dig th' Immortall Fount,
> Take th' *Holy-Ghost,* typ't in a *Siluer Doue.*[34]

One wonders if Du Bartas' Urania is some mixture of classical muse, astrological divinity, human wit and the Holy Ghost. The identity and function of Milton's muse is no less challenging to settle, but his emphasis clearly shifts even more toward Christianization.

Milton invokes Urania as his muse in Book VII of *Paradise Lost*: 'Descend from Heav'n Urania', he petitions. However, he immediately indicates apprehension and calls for qualification:

> by that name
> If rightly thou art call'd, whose Voice divine
> Following, above th' Olympian hill I soare,
> Above the flight of Pegasean wing.
> The meaning, not the Name I call: for thou
> Nor of the Muses nine, nor on the top
> Of old Olympus dwell'st, but Heav'nlie borne,
> Before the Hills appeerd, or Fountain flow'd,
> Thou with Eternal Wisdom didst converse,
> Wisdom thy Sister, and with her didst play
> In presence of th' Almightie Father, pleas'd
> With thy Celestial Song.[35]

The distancing of 'if rightly thou art called', and the negations of 'the meaning not the Name' and 'Nor of the Muses nine' quite clearly indicate

---

[33] Du Bartas, *The Divine Weeks and Works of Guillaume Du Bartas*, 1.2.1101-102
[34] Du Bartas, *The Divine Weeks and Works of Guillaume Du Bartas*, *L'Uranie* Stanza 62.
[35] Milton, *Complete Poems and Major Prose*, 7.1–12

Milton's distinguishing his provisionally employed *topos* from classical usages. He also references the story of Orpheus being torn apart by the followers of Bacchus, and labels the muse Calliope 'an empty dreame' for being unable to protect her son.[36] In doing so, the ontological status of all the muses is called into question. According to Gregory, this invocation is an instance of 'Milton's consistent habit of using myth to suggest truth while stressing its inadequacy for such a task'.[37] Milton appears to be interested in something concerning the 'meaning' of Urania as conditioned by her associative evolution in history, while proffering 'a clear directive to the reader to avert the eyes from Urania's classical origins'.[38] Consider that 'Wisdom thy sister' is a clear allusion to the Hebraic figure of Proverbs 8:22–25:

> The LORD possessed me in the beginning of his way, before his works of old.
> I was set up from everlasting, from the beginning, or ever the earth was.
> When there were no depths, I was brought forth; when there were no fountains abounding with water.
> Before the mountains were settled, before the hills was I brought forth.[39]

We have noted earlier that the muses were born after the creation of the world, on Mount Olympus (their favourite haunt was near fountains), through Zeus and Memory; but Milton's figure antedates them; she *is not one of them.*

Once Milton asserts that he calls on 'the meaning, not the name' of Urania, the question arises: 'whom *is* Milton invoking?' When Milton invokes the Muse in Book I (without naming her), he immediately conflates her with the Spirit who created the world and who inspired Moses, and then with the Dove. An invocation also takes place in Book III in which the Father and 'Holy Light' are called upon to inspire the poet;

---

[36] Milton, *Complete Poems and Major Prose*, 7.32-38, 7.39.

[37] Gregory, *Milton and the Muses*, p. 96. Gregory points to other episodes in *Paradise Lost* such as Milton's provisional comparison of Mulciber to the architect of pandemonium (1.738-47), and the episodes in Eden in which the fruit with golden rind is compared to 'Hesperian Fables true, / If true, here only' (4.249-50).

[38] Stevie Davies and William Hunter, 'Milton's Urania: The meaning, not the name I call', *SEL: Studies in English Literature, 1500-1900* 28, no. 1 (1988): p. 100.

[39] *The Holy Bible,* Authorized King James Version (Grand Rapids, MI: Zondervan, 1995), Proverbs 8:22–25.

Book VII, as we have seen, calls upon a 'meaning' of Urania, who is the sister of Eternal Wisdom; and Book IX calls upon a 'Celestial Patroness'. Scholars have tried to make sense of these distinct invocations, and their answers have usually tried to identify Milton's muse with a member of the Christian trinity.[40] Their disagreements about identity are worth reviewing here, because they corroborate with my point that Milton's Urania is not the celestial intelligence of the starry sphere. William Hunter argues for Urania's identity as the *Logos* of God, based partly on the fact that Jesus is almost certainly invoked in Book III, inferring that Milton invokes the same being in both places. Moreover, Hunter does some background reading into Neo-Platonic understandings of Urania, including Milton's contemporary Ralph Cudworth, and concludes that both Urania and Athena, as mind or intellect, 'may represent the same second stage of the Neoplatonic trinity'.[41] This implies that Urania could, by analogy, be understood by Milton's readers as the second member of the Christian Trinity. John Steadman argues for Urania's identity as 'a poetical personification of an attribute of the Father', given her kinship with biblical Wisdom who also appears to be one: 'If so, then the most likely probability is that she represents the original harmony existing in the divine mind, reflected in the divine decrees, and subsequently realized in the 'pulcherrimo ordine' of the visible and invisible worlds'.[42] Hunter later modified his view, and in an article with Stevie Davies, proposes that Book VII refers to the Spirit of God. They explain,

> although the Spirit 'cannot be a God' in Milton's view and accordingly cannot be 'an object of invocation' (VI: 295) it is the vehicle for appeals to the Father and for the Father's responses, just as is the Son, 'in whose name' Christians pray. Thus the poet prays to the Spirit as Urania here as he had prayed to the Son as light in Book III, not as the finally divine and responsive God but as

---

[40] Another important school of thought has been the scholars who consider Urania to express some aspect of art or beauty. Helen Gardner, *A Reading of Paradise Lost*, considers Urania the 'poetic embodiment of Milton's belief in his vocation' (Oxford: Clarendon, 1965), p. 20. James Holly Hanford, *John Milton, Englishman* asserts her to be 'the voice of Biblical revelation . . . and the creative power itself' (New York: Crown, 1949), pp. 179-80. Gregory, *Milton and the Muses* states that Urania is 'God's wisdom in the form of art, specifically religious poetry' (p. 120).
[41] William B. Hunter, Jr, 'Milton's Urania', *SEL: Studies in English Literature, 1500-1900* 4, no.1 (1964): p. 38.
[42] John M. Steadman, *Milton's Biblical and Classical Imagery* (Pittsburgh, PA: Duquesne University Press, 1984), p. 67.

the medium of communication with the Father and specifically as the vehicle through which the ineffable Father bestows his Grace upon an individual.[43]

Treip agrees that the multi-referentiality of the invocations works best with the multi-referentiality that Milton assigns to the Holy Spirit in *De Doctrina Christiana*.[44] David Ainsworth concurs: Urania is to 'find' the audience, the community of believers, and therefore is the Holy Spirit.[45] In the variegated debate about the identity of Milton's muse, the one matter all scholars agree upon is that she is *not* the Urania of classical mythology or the astrological divinity of Neo-Platonism.

Milton is allegorizing, 'other-speaking' in the most radical sense: using the ore of mythology to bring out the gold of Christian truth, seeking out the meaning beyond the name. Indeed, *Paradise Regained* simply invokes 'Thou Spirit',[46] leaving behind the former, classical part of the allegorical equation. A recent article has proposed that Milton was more or less hostile to allegory, based on broader patterns of Protestant antipathy toward this mode of interpretation: Vladimir Brljak argues that the character of Satan in *Paradise Regained* presents allegorical 'either/or' options for interpreting the literal and figurative, whereas Jesus the authentic hermeneut reads 'both/and' in the Old Testament for a typological reading. I strongly believe that Milton was open to allegorical readings of the classics, and his invocation to Urania would have to be intensely 'satanic' in Brljak's sense: Milton aims for an allegorical 'either/or' figure of Urania, dissociating the meaning from Urania's putative reality.[47]

We turn at last to the issue of cosmology in *Paradise Lost*. The meaning of Urania in the Ptolemaic cosmos had much to do with her role and place within it. With Karen Edwards, Malabika Sarkar, John Leonard and Dennis Danielson, I concur that Milton's cosmos is strongly influenced by the new astronomy and is not represented in a Ptolemaic manner, in spite of the long tradition that perceived it otherwise. This wave of current scholarship

---

[43] Hunter, 'Milton's Urania', p. 100.

[44] Treip, *Descend*, outlines that for Milton 'spirit' can mean the Father, Christ as Word, generative power, gifts conferred on individuals, light, the light of truth, and a divine impulse (pp. 50–51).

[45] David Ainsworth, 'Getting Past the Ellipsis: The Spirit and Urania in *Paradise Lost*', in *Renaissance Papers 2012*, eds. Andrew Eric Shiflett and Edward Gieskes (New York: Camden, 2012), pp. 117–25.

[46] Milton, *Complete Poems and Major Prose*, 1.8.

[47] Vladimir Brljak, 'The Satanic 'Or': Milton and Protestant Anti-Allegorism', in *Review of English Studies* 66 (2015): pp. 403–422.

helps to elucidate the implications of Milton's Urania. Hell is at some inestimable depth below the cosmos, not at the centre of the earth; outside of Milton's Stygian abyss is the enigmatic realm of Chaos, which is also outside of our cosmos – and does not exist in the Ptolemaic cosmos. Satan emerges out of Chaos and the Paradise of Fools to approach the gates of heaven, but only then is able to observe our cosmos as a separate creation, whereas the Ptolemaic cosmos is a continuous scheme up to the Empyrean. Milton's 'Empyrean,' as Edwards demonstrates, is uncircumscribed, more conceptual than spatial, and terms like 'height', 'dwell' and 'place' cannot 'have anything like ordinary human meaning'.[48] Satan's flight changes from vexed and difficult in Chaos, to winding 'with ease / Through the pure marble Air his oblique way'.[49] This kind of flight is not Aristotelian, as a perusal of Robert Burton's 'Digression of Aire' indicates: '[heaven] is not hard and impenetrable, as Peripateticks hold, transparent, of a *quinta essentia*, but that it is penetrable & soft as the air itself is, & that the Planets move in it, as Birds in the Air, Fishes in the Sea.'[50] Moreover, whereas the *Cosmographia* constantly refers to the zodiac houses as visible in the heavens, affecting cosmological affairs, and as markers of place to which Natura travels, Milton's poem refers to two zodiac houses in order to indicate the breadth of Satan's survey of the world, and later to explain the axial shift accounting for seasons.[51] It is fairly clear that the cosmos of *Paradise Lost* is not vertiginous, as the Ptolemaic cosmos is: Uriel points to a 'neighbouring moon' beside the earth, and the narrator records Satan's disorientation within the three-dimensional plane: 'but up or downe / By center, or eccentric, hard to tell, / Or Longitude'.[52] Significantly, the golden chain of the *Cosmographia* links the sphere of the moon to the earth, whereas in *Paradise Lost*, the chain links heaven to the 'pendant world',[53] implying that nature, for Milton, encompasses the entire created cosmos.

---

[48] Edwards, 'Cosmology', p. 120.

[49] Milton, *Complete Poems and Major Prose*, 3.563–64.

[50] Robert Burton, *The Anatomy Of Melancholy, What It Is. With All The Kindes, Cavses, Symptomes, Prognostickes, And Severall Cvres Of It. In Three Maine Partitions with Their Seuerall Sections, Members, and Svbsections. Philosophically, Medicinally, Historically, Opened And Cvt Vp.* (Oxford: Printed by Iohn Lichfield and Iames Short, for Henry Cripps, 1621), p. 324.

[51] Milton, *Complete Poems and Major Prose*, 3.557–60, 10.668–79.

[52] Milton, *Complete Poems and Major Prose*, 3.726, 3.574-76.

[53] Milton, *Complete Poems and Major Prose*, 2.1052.

Regarding the nature of earth, Milton presents a highly 'Galilean' picture. Aristotle had proclaimed the earth to be the lower part of the cosmos, constituted by heavy elements that did not have it in their nature to shine. That Milton depicts earth as emitting light 'though but reflected', is very much an anti-Aristotelian doctrine with far-reaching consequences that concern the possibility of earth being a star itself and emitting light and influence.[54] In Galileo's *Dialogue Concerning the Two Chief World Systems* (1632), his Copernican protagonist Salviati has a long section on reflected light, and excitedly concludes with his belief in the 'beautiful resemblance' between earth and other planets: 'if it is true that the planets act reciprocally upon the earth by their motion and by their light, perhaps the earth is no less potent in acting upon them by its own light – and possibly by its motion, too'.[55] In Galileo's *Sidereus Nuncius*, his report on early telescopic observations of the moon and the discovery of moons around Jupiter, he proclaims that he will refute those who ban earth from joining the 'dance of the stars,' and that he will do so by demonstrating that the earth is 'movable and surpasses the moon in brightness, and that she is not the dump heap of filth and dregs of the universe'.[56] In a similar vein, Milton describes Edenic earth-dwellers as 'perfect, not immutable',[57] and exalts the habitation of humankind in 8.91-94:

> […] the Earth
> Though, in comparison of Heav'n, so small,
> Nor glistering, may of solid good containe
> More plenty then the Sun that barren shines.

This passage is at striking odds with the Ptolemaic world-picture of the heavy, ignobly mutable, bottom-rung, excrementary earth.[58]

---

[54] Milton, *Complete Poems and Major Prose*, 3.723.

[55] Galileo Galilei, *Dialogue Concerning the Two Chief World Systems*, trans. Stillman Drake (Berkeley, CA: University of California, 1953), p. 110. Of course, this reality was not at odds with astrological belief *per se*, but it did pose questions about the old system of astrology associated with Urania.

[56] Galileo Galilei, *Sidereus Nuncius*, trans. Albert Van Helden (Chicago, IL: University of Chicago, 1989), p. 57.

[57] Milton, *Complete Poems and Major Prose*, 5.524.

[58] Salviati, in the *Dialogue*, states something that sounds similar to Milton: 'we plainly see and feel that all generations, changes, etc. that occur on earth either directly or indirectly designed for the use, comfort, and benefit of man' (p. 69).

Perhaps most significant is the 'simple, startling truth' that Leonard notes: 'there are no spheres in *Paradise Lost*'.[59] Indeed, whereas in the *Cosmographia*, Urania's ascent and descent through the celestial spheres involves weariness, constant obstacles and narrative benchmarks, *Paradise Lost* depicts flight through the cosmos as through a sort of homogeneous space. The only place there is any sphere, is in the satirical Paradise of Fools, whose inhabitants 'pass the Planets seven, and pass the fixt, / And that Crystalline Sphear whose ballance weighs / The Trepidation talkt, and that first mov'd'.[60] We might add that Satan does not encounter a Ptolemaic sphere of fire or of air as he enters the earth.

Milton has left Urania's home, the sphere of the fixed stars (and in some accounts the crystalline sphere), in ruins: it has no place in the new cosmology Milton has painted. Her jurisdiction would have to be expanded erratically, given the galaxy 'powdered with stars', the 'innumerable Starrs', the stars 'thick as a field',[61] the description of which, Danielson explains, relates to Galileo's telescopic revelation that made the universe to appear 'more and more as an array of discrete bits',[62] with many potential centres, not finite and uniform along a uniform sphere with a central earth. Recalling Figure 1's portrayal of the spheres, there were proportions of distance between the heavenly spheres, and these corresponded to tonal distances in the musical octave. Agrippa, in fact, writes that 'there are some who find out the harmony of the Heavens by their distance one from another. For that space which is betwixt the Earth and the Moon, *viz* an hundred and twenty six thousand *Italian* Miles, maketh the Intervall of a Tone,' and points out that the distance 'from [Saturn] to the starry firmament is […] the space of an half Tone'.[63] Milton's 'array of discrete bits' problematizes the exact proportions requisite for what people supposed would account for the muses' harmonious music. Accordingly, in a cosmos of shattered spheres, and other possible suns and worlds, Urania does not retain her place in a hierarchy of influence among the muses or intelligences, which, in fact, do not seem to exist for Milton.

Many of the astrological images that exist in *Paradise Lost* are wrapped up in Satan: the flaming figure 'whose stature reached the sky', the one who, 'as from a cloud his fulgent head / And shape star bright appeared'.[64]

---

[59] Leonard, *Faithful Labourers*, p. 709.

[60] Milton, *Complete Poems and Major Prose*, 3.481–83.

[61] Milton, *Complete Poems and Major Prose*, 7.581, 3.565, 7.358.

[62] Danielson, *Paradise Lost and the Cosmological Revolution*, p. 88.

[63] Agrippa, *Three Books*, p. 261

[64] Milton, *Complete Poems and Major Prose*, 4.988, 10.449–50.

When Sin and Death meet him, 'Satan in likeness of an angel bright / Betwixt the Centaur and the Scorpion steering / His zenith, while the sun in Aries rose'.[65] Perhaps most important is the passage in Book II when Satan stands

> Unterrified, and like a comet burned,
> That fires the length of Ophiucus huge
> In the Arctic sky, and from his horrid hair
> Shakes pestilence and war.[66]

Sarkar convincingly argues that Satan's appearances in this specific place relate him clearly to the Ophiucus comet of 1618, and reflect Milton's 'fierce indictment of the false hopes raised of the immediate advent of the promised millennium and the misreadings of celestial signs that fuelled such hopes'.[67] Accordingly, Milton writes wittingly with astrological signs and images, drawing upon the currency of astrological-historical beliefs and events. The tying of Satan to comets and stars, however, is not evidence that he subscribed to the truth of astrological tenets; in fact, it is quite likely that he was critiquing contemporary prognosticators.

There are no astrological divinities in Milton's universe: in the *Cosmographia*, the planets are actively affecting the affairs of the universe below them, and are interactive, while the angelic beings throughout the cosmos are silent and merely observed as bystanders. In *Paradise Lost*, it is the exact reverse: the planets, by and large, move silently while the angelic beings act, converse, and sing. Admittedly, the moon, the sun, and the planets (in a brief moment of personification) do worship God 'In mystic Dance not without Song', and heavenly bodies do affect the earth in some way relating to 'nourishment'; Satan states that in earth, 'all [the planets'] known vertue appeers'; the Pleiades, too, shed 'sweet influence'.[68] Perhaps most significantly, after the fall, God mandates the angels to make the 'planetary motions and aspects / In Sextile, Square, Trine, and Opposite, / Of noxious efficacie'.[69] Here is a relatively strong statement about planetary influence, even though there is no sense that the planets have distinct personalities and divine attributes. D. C. Allen's learned survey of Renaissance astrology concludes that even 'the opponents of astrology

---

[65] Milton, *Complete Poems and Major Prose*, 10.327–29.
[66] Milton, *Complete Poems and Major Prose*, 2.708–11.
[67] Sarkar, *Cosmos and Character in Paradise Lost*, p. 122.
[68] Milton, *Complete Poems and Major Prose*, 5.178, 5.421–26, 9.110, 7.375.
[69] Milton, *Complete Poems and Major Prose*, 10.658–60.

admitted that the stars had an influence on man, but insisted that it was impossible to determine what the influence was'.[70] Allen's statement concerns judicial astrology: the branch of the discipline which sought out answers to all manner of issues, from natal predispositions to forecasts of destiny, to fortuitous world events. Only full-fledged astrologers held this belief, and it was very seriously and systematically doubted by Milton's time. However, Milton seems to have retained the belief that there were four elements 'earth, flood, air, fire', and four elementary qualities 'hot, cold, moist, and dry'.[71] These are the same elements that Aristotle discusses in his several of his writings on nature: "The elementary qualities will be four: hot with dry and moist with hot, and again cold with dry and cold with moist. And these four couples have attached themselves to the apparently 'simple' bodies (Fire, Air, Water, and Earth)".[72] Keith Thomas explains that of all the astrological doctrines, it was most 'generally accepted that the four elements constituting the sublunary region – earth, air, fire and water – were kept in their state of ceaseless permutation by the movement of the heavenly bodies'.[73] Therefore, although it is reasonable to conclude that Milton accepted the most basic premise of astrology, he did not buy into the entire cosmos-picture associated with the older versions of it, and almost certainly not judicial astrology.

Not surprisingly, then, Urania has limited work to do in such a cosmos. Her job of conducting the music of the spheres becomes the 'Harmonious sound / On golden Hinges moving, to let forth' brought by 'The King of Glorie in his powerful Word / And Spirit coming to create new Worlds'.[74] The angels also function as prominent music makers in Book III as in Book VII. Her work of creating and sustaining the intellective and spiritual capacities of mankind are done by the Son directly; Milton, moreover does not suggest that man's material and spiritual creation are separate affairs. The invocation to Urania does appropriately preface a discourse on Astronomy. However, it is a discourse that would be unthinkable for Capella's Urania (who counsels incessant study of the heavens and its secrets), since Raphael admonishes Adam not to concern himself over the

---

[70] D. C. Allen, *Star-crossed Renaissance: the Quarrel about Astrology and Its Influence in England* (New York: Octagon, 1966), p. 245.

[71] Milton, *Complete Poems and Major Prose*, 3.715, 2.898.

[72] Aristotle, *On Generation and Corruption* in *Delphi Complete Works of Aristotle*, trans. H. H. Joachim (Hastings: Delphi, 2013), II.iii.

[73] Keith Thomas, *Religion and the Decline of Magic* (New York: Scribner, 1971), p. 285.

[74] Milton, *Complete Poems and Major Prose*, 7.206-209.

movement of the heavens and to think on 'onely what concernes thee and thy being'.[75] This view is unacceptable to any strong astrological doctrine, which states that the movement and operations of the heavens have everything to do with human *being*. Raphael ridicules the enterprise of those who 'build, unbuild, contrive' the heavens, suggesting the artificiality of cosmos-building at large.[76] The Urania of the centuries-old, stable cosmos could never support such an evicting discourse for her own habitation. In fact, as discussed earlier, Urania is not an inhabitant here at all. Her name is invoked for its meaning, not its mettle.

When one considers all these matters of a new cosmology in *Paradise Lost* and stumbles across Milton's remark that inspired poetry is not 'to be obtained by the invocation of Dame Memory and her Siren daughters, but by devout prayer to that eternal Spirit who can enrich with all utterance and knowledge, and sends out his seraphim with the hallowed fire of his altar, to touch and purify the lips of whom he pleases',[77] does it behove one to conclude that Milton has trivialized the muse – that scholars must believe a 'descent' in Urania's Miltonic importance? In many ways, I think not. He *does* invoke a siren daughter, an epic invocation complete with the word 'muse,' risking the poetico-mythic-astrological name of 'Urania'. This has a very significant function, which Gregory expresses adroitly: 'In their role as inspiratresses of the arts, the Muses were agents external to the poet that enabled him to achieve a level of excellence in his work inexplicable in terms of his own abilities'.[78] This is a conventional and noble, if self-authorizing literary move, even as Milton removes the identity of his celestial patroness to something more foundationally Judeo-Christian. Moreover, it seems plausible that Milton was interested in Urania's association with contemplation, astronomy, and music. However, I find it implausible to state, like Philip Edward Philips does, that Milton is syncretistic. Rather, he is eager to transform and simultaneously extricate the meaning of this astrological divinity, while retaining some of her trappings, both in his invocation and in his cosmos.

---

[75] Milton, *Complete Poems and Major Prose*, 8.174.

[76] Milton, *Complete Poems and Major Prose,* 8.81.

[77] Milton, 'Reason of Church Government', p. 671.

[78] Gregory, *Milton and the Muses*, p. 63.

Ilm-i nudjum and
Eighteenth-Century Ottoman Court Politics

R. Hakan Kirkoğlu

Abstract: Astrologers (*munajjims*) who were responsible for preparing calendars and calculating prayer times secured a bureaucratic position in the Ottoman court from the times of Bayezid II (1481-1512). In the early modern Muslim world, an astrologer was the time expert whose time concept was symbolic, eschatological and millennial. The yearly predictive works (*ahkams*), produced and submitted to the sultan and the ruling elite every Spring (*Newruz*), were intended to serve as political guidelines as well as an agenda for court members. In their yearly judgements, astrologers often highlighted sensibilities of the sultan and refrained from using expressions which could create potential unrest in different layers of his administration. In this study, I investigated the works of eighteenth century chief-astrologer Halil Efendi who served under three sultans: Mahmut I (1730-1754), Osman III (1754-1757) and Mustafa III (1757-1774). Halil Efendi was a companion to Mustafa III, and his yearly predictions reveal how he was influential in court politics before the war against the Russians (1768-1774).

Astrology in Ottoman Historiography

Even though astrology was labeled a 'wretched' subject by the late historian of science, George Sarton, the history of astrology as an important element of ancient, medieval and early modern science and culture has received much scholarly attention in recent decades.[1] Historians of thought like Anthony Grafton, Ann Geneva, Otto Neugebauer and Tamsyn Barton, as well as historians of political culture like Cornell Fleischer and Azfar Moin, have explored different aspects of this most neglected subject, including its social, political, religious and scientific dimensions.[2]

---

[1] George Sarton, 'The Book of the Zodiac', *Isis* 41 (1950): p. 374.

[2] Anthony Grafton, 'Starry Messengers: Recent Work in the History of Western Astrology', *Perspectives on Science (the Massachusetts Institute of Technology)* 8, no. 1 (2000); Anthony Grafton, *Cardano's Cosmos, The Worlds and Works of a Renaissance Astrologer* (Cambridge: Harvard University, 1999); William R. Newman and Anthony Grafton, eds, *Secrets of Nature, Astrology and Alchemy in Early Modern Europe* (Cambridge: The MIT Press, 2006); Ann Geneva, *Astrology*

R. Hakan Kirkoğlu, '*Ilm-i nudjum* and 18th century Ottoman Court Politics', *Culture and Cosmos,* Vol. 18, no. 2, Spring/Summer 2014, pp. 125-141.
www.CultureAndCosmos.org

Despite a number of pioneering studies by Cornell Fleischer, nevertheless astrology—or *Ilm-i nudjum* in the Ottoman context—still remains an understudied topic. My aim in this context is to examine *Ilm-i nudjum*'s place in Ottoman court politics in the eighteenth century. Astrology was used as a tool of political authority, stability, legitimacy and efficient propaganda apart from its supposedly predictive value.

In this perspective, my purposeful selection of court astrologer Fethiyeli Halil Efendi as a subject is firstly due to his very long tenure as a chief-*munajjim*, serving under three sultans—Mahmut I, Osman III and Mustafa III. Indeed, such a long continuity coincides with a relatively peaceful period in Ottoman history. Moreover, I presume that Sultan Mustafa III's personal interest in *Ilm-i nudjum*, which is a well-documented fact, had a major role in this respect.

Another interesting fact is that, due to Mustafa III's special interest in *Ilm-i nudjum*, there was a notable increase in the number of translated astrological works which were dedicated in his name. According to the list of these dedicated works by IRCICA, 13 of 106 belong to Mustafa III, 14 to Bayezid II and 9 to Suleiman I, while the rest are sparsely distributed, reflecting increased activity in the production of astrological works in Mustafa III's time[3].

In the twentieth century, mainstream works on astrology that have been written under a rigid type of Enlightenment thinking usually tend to look at the subject in terms of a battle between progress and inertia.[4] In most of the histories of science, as there has been no any other available ground to speak in contrariwise in Academia, *Ilm-i nudjum* has been classified as a residual form of ancient knowledge. To give an example, when Adnan

---

And The Seventeenth Century Mind: William Lilly And The Language Of The Stars (Manchester: Manchester University Press, 1995); Otto Neugebauer, 'The Study of Wretched Subjects', *Isis* 42 (1951): p. 111, reprinted in O. Neugebauer, *Astronomy and History: Selected Essays* (New York: Springer-Verlag, 1983); Tamsyn S. Barton, *Ancient Astrology* (London: Routledge, 1994); Tamsyn S. Barton, *Power And Knowledge: Astrology, Physiognomics, and Medicine under the Roman Empire* (Michigan: The University of Michigan Press, 2002); Cornell Fleischer, *Bureaucrat and Intellectual in the Ottoman Empire: The Historian Mustafa Ali (1541-1600)* (Princeton: Princeton University Press, 1986); A. Azfar Moin, *The Millennial Sovereign, Sacred Kingship and Sainthood in Islam* (New York: Columbia University Press, 2012).
[3] Ekmeleddin Ihsanoğlu, ed., *Osmanli Astroloji Literatürü Tarihi ve Osmanli Astronomi Literatürü Tarihi Zeyli*, Vol.1 (Istanbul: IRCICA, 2011), p. 72.
[4] Sarton, 'The Book of the Zodiac', p. 374.

Adıvar mentions Kadızade Rumi, citing *Asar-ı Bakiye* by Salih Zeki (1864-1921, founder of Astronomy department of Istanbul University), he points out with praise that Kadızade Rumi never wrote a single sentence on *Ilm-i nudjum*.[5] The contemporary historian of science, Yavuz Unat, touches on astrology very sparsely in its relation to Ptolemy's *Tetrabiblos*, stating that astrology was a scientific religion in the age of Ptolemy.[6] Particularly, we need to note that scientific progress based on observation, experimentation and a secular worldview led to the demise of astrology, which had been constructed on a paradigm of divine correspondences in the old enchanted world. On the other hand, we should also remember that *Ilm-i nudjum* has been attacked by both religion and science. As a by-product of these major norms, *Ilm-i nudjum* and other related subjects of the occult in general have been relatively shunned subjects in Ottoman studies.

Until recently, another factor behind the neglect of astrology has been the lack of a proper catalog of related manuscripts. Fortunately, this situation has now changed with the publication of Ottoman astronomical and astrological texts by a team headed by Günay Kut from Boğaziçi University in 2007.[7] Still, to my knowledge, Ottoman astrology has to date been the subject of only two scholarly works: one by Salim Aydüz on the establishment of court astrologers in the Ottoman Empire and another by Gülçin Tunalı Koç on a provincial astrologer's life as a micro-history.[8] In addition, Ekmelleddin İhsanoğlu's statement, in his preface to the IRCICA's *History of Ottoman Astrology Literature*, that astrology has an

---

[5] Adnan Adivar, *Osmanlı Türklerinde İlim* (Istanbul: Remzi Kitabevi, 1982), p. 20.
[6] Yavuz Unat, *İlkçağlardan Günümüze Astronomi Tarihi* (Istanbul: Nobel Yayınevi, 2001), pp. 56-57.
[7] Günay Kut, *Kandilli Rasathanesi El yazmaları 1: Türkçe yazmalar* (Istanbul: Boğaziçi Üniversitesi Yayınları, 2007); Günay Kut, *Kandilli Rasathanesi El yazmaları 2: Arapça-Farsça yazmalar* (Istanbul: Boğaziçi Üniversitesi Yayınları, 2012).
[8] Salim Aydüz, 'Osmanlı Devletinde Müneccimbaşılık ve Müneccimbaşılar' (master's thesis, Istanbul University, 1993). Ayduz's thesis, as a first attempt to explain *Ilm-i Nücûm*'s role in the court tradition, provides chronological and institutional content based on biographies of *munajjim*s and archive sources. Gülçin Tunalı Koç, 'Sadullah el-Ankaravi: Daily concerns of an Ottoman Astrologer' (master's thesis, Boğaziçi University, 2002). On the other hand, with its specific narrative, Koç's thesis sheds light on the life and daily concerns of an Ottoman *munajjim*, Müderriszâde Sadullah el-Ankaravi.

important place among the bricks that have built the edifice of science, I presume indicates a new awareness. [9]

Definition of *Ilm-i nudjum* in Islamic sources: its classification and relation to *Ilm-i Felek*, astronomy

At this point I would like to draw attention to the relationship between astrology and astronomy, as both disciplines share common historical grounds but also fierce disputes. In the Arabic language, the technical term for astrology, *ilm* (or *sina'at*) *ahkam an-nudjum*, means the science (or art) of the decrees of the stars. Until the ninetheenth century, the distinction between astrologer (*munajjim*) and astronomer (*falaki*) was not indicated in the Arabic language. In Islamic civilization, astronomy (*ilm-el felek*) and astrology (*Ilm-i nudjum*) were regarded as sister sciences, having complementary qualities, since according to Ptolemy in his *Tetrabiblos* the science of the stars consists of two parts.

Astrology is comprised of two main branches, natural and judicial astrology. [10] Natural astrology consists of the observation of the influences of the stars on the natural elements while judicial astrology dwells on the stellar influences upon human destiny. While natural astrology was usually received and accepted as a beneficial practice, judicial astrology purportedly was seen as an intervention into God's will and was therefore strongly opposed by theologians. Furthermore, astrology as a body of knowledge in Islam has been divided into principal categories of application in which several divisions were not fully esteemed or rather not admitted due to their disapproval by the Ptolemaic/Aristotelian model of thinking. [11]

Courtly practice of astrology

Despite its controversial nature, *Ilm-i nudjum* was widely practiced by Muslim astronomers, *munajjims*. Its use was not only limited to the elite classes or to court life; certain people who desired to win the favor of a ruler resorted to astrology by casting charts and making predictions. [12]

---

[9] İhsanoğlu, *History of Ottoman Astrology Literature,* p. XXVII.

[10] T. Fahd, *Ahkâm al-Nudjum*, Encyclopedié de l'Islam, Nouvelle Edition VIII, (Leiden, 1995), pp. 108-111.

[11] Carlo Nallino, *Sun, Moon, and Stars*, Encyclopedia of Religion and Ethics, Vol. 12, p. 89.

[12] George Saliba, 'The Role of the Astrologer in Medieval Islamic Society', *Magic and Divination in Early Islam*, ed. Emilie Savage-Smith (Ashgate, Variorum, 2004), p. 357.

Although *Ilm-i nudjum* was criticized by the majority of Muslim religious authorities, a margin was left for it in view of its perceived usefulness in the state bureaucracy. Finding and setting dates for certain events, festivals and courtly matters through *Ilm-i nudjum* could inherently be seen to provide order and stability for the state hierarchy in cosmological terms. The Ottoman state archives contain numerous records of astrological practice carried out by chief-*munajjim*s. These records refer to administrative correspondence and other items, such as the ground-breaking ceremony of *Dâr'ül Aceze* (Poorhouse) (1892),[13] the wedding ceremony of Abdulhamit I's daughter Esma Sultan,[14] the date of a banquet during the reign of Ahmed III (8 April 1731),[15] written documents of geomancy (7 November 1745),[16] records of yearly payments and daily wages to *munajjim*s, a *munajjim*'s notice of a propitious moment for dispatching a standard (*sancak-ı şerif*) (21 August 1789),[17] flying a flag on a warship (10 August 1790),[18] and a notice indicating the date of a campaign (28 June 1794),[19] to name just a few. Doubtless, such records reveal how *munajjim*s were used to participate in administrative processes as timing experts. In the minds of the people, the selection of a certain date and hour created an aura of immovability, power and legitimacy for the Sultan's actions and the state itself. Stephen P. Blake states that, in the early modern Islamic world, the *munajjim* was the time expert whose time concept was symbolic, eschatological and millennial.[20] The resiliency of *Ilm-i nudjum* rests in its strong relation with divinity, whose perfect order and cyclical nature were believed to be reflected in planetary movements. Hence, similar to the traditional role of court astronomers/astrologers in the West, Muslim astronomers/astrologers in the courts of the Safavid and Mughal Empires were employed for compiling planetary tables (*zic*s), writing almanacs, predicting future prospects like battles and marriages, and casting the ruler's horoscope.

---

[13] *Yıldız Mütenevvi Maruzat Evrakı*, file no. 70, folder no . 95.

[14] *Topkapı Sarayı*, folder no. 9830.

[15] *Cevdet Saray*, file no. 37, folder no. 1873.

[16] *Cevdet Maarif*, file no. 77, folder no. 3801.

[17] *Hatt-ı Hümayun tasnifi*, file no. 1378, folder no. 54248.

[18] *Hatt-ı Hümayun tasnifi*, file no. 1392, folder no. 55613.

[19] *Hatt-ı Hümayun tasnifi*, file no. 237, folder no. 13160.

[20] Stephen P. Blake, *Time in Early Modern Islam: Calendar, Ceremony, and Chronology in the Safavid, Mughal and Ottoman Empires* (New York: Cambridge University Press, 2013), p. viii.

Akin to other Islamic empires, the practice of *Ilm-i nudjum* can be seen as a part of state bureaucracy. Although the first calendars as court almanacs were prepared during Murad II's reign (1421-1451), the names of these calendar makers are still unknown to us. However, we know that Ali Kuşçu (d. 1474) was invited by Mehmed II to teach mathematics and astronomy in various madrasas in Istanbul.[21] Salim Aydüz states that astrological activities were apparently on the increase during the reign of Bayezid II. Andrea Gritti, later doge of Venice, states that Bayezid II was a great connoisseur of astrology and theology.[22] The first admission of astrologers to the Ottoman court was in the time of Bayezid II, as seen from the record books of state rolls dated between the years 1537 and 1544 in the Topkapı Palace.[23] Ismail Hakkı Uzunçarşılı writes that 'the astrologers were to determine the auspicious times (*eşref saati*) for the ceremonies of accession of the sultans, the births of sultan's sons, waging wars and the movement of the armies, concessions of the seals to the viziers, launching of the galleys, sultan's feasts, weddings, etc.[24] The sultans and the dignitaries were tending to and practicing these special moments for their fortunes customarily. For this reason, the custom of astrology was seen as an important practice in the state affairs'. [25]

Chief-*munajjims* belonged to the outer circle of the Ottoman Palace, '*birun*' circle and they were supposed to complete typical madrasa education along with special tuition in *Ilm-i nudjum* and calendar making from learned people who had an interest in this direction. Chief-*munajjims* themselves as members of the *Ilmiye* class could be employed as tutors as well. They could also take office as professors (*mudarrises*) or could be assigned as *kadı*s or even as chief physicians (*hassa hekim*). Time-keeping houses (*muvakkithane*s) were one of these places where students of *Ilm-i*

---

[21] Salih Zeki, *Asar-ı Bakiye* (Ankara: Babil, 2004), pp. 133-139.

[22] E. S. Creasy, *History of the Ottoman Turks: from the beginning of their Empire to the present Time. Chiefly founded on Von Hammer*, Vol. VI (London: Richard Bentley, 1854), p. 625.

[23] Ö. Lütfü Barkan, 'H.933-934 Tarihli Bütçe Cetveli ve Ekleri (İlave V)', *İktisat Fakültesi Mecmuası* XV, nos. 1-4 (Istanbul: 1955), pp. 314-329.

[24] Aydüz, in his master's thesis 'Osmanlı Devletinde Müneccimbaşılık ve Müneccimbaşılar', provides a list of the 37 chief-*munajjims* in the Ottoman court since the sixteenth century.

[25] Ismail Hakkı Uzunçarşılı, *Saray Teşkilatı* (Ankara: Türk Tarih Kurumu, 1988), p. 269.

nudjum were actively involved. The Fatih, Yavuz Selim and Şehzade mosques' time-keeping houses were well-known places in this regard.[26]

Although certain educational traits and degrees were mentioned towards attaining chief-astrologership, especially in the 19th century, recommendation by the chief-physician and familial connections were an important route in acquiring such a position in the court. Their selection and assignment procedures involved bureaucratic lines and the chief physician proposal was crucial.

According to Salim Ayduz, a total of 37 chief-*munajjims* served in the Ottoman court during and after the reign of Bayezid II (1481 to 1512).[27] Their length of service varied depending on the conditions of the Sultan in power. While some of them were dismissed upon the accession of a new sultan, others were discarded due to their miscalculations and mistakes in calendar making. In only one case was a *munajjim* executed: Huseyin Efendi in 1650 (during the reign of Mehmed IV) for reasons of enmity and corruption.

Despite *Ilm-i nudjum*'s fragile credibility and its contested nature, the chief-*munajjims'* *Ilmiye* status would have bestowed on them a certain positional power in speaking out their point of view. Their power could grow or decline depending on the personal affinity and special interest of sultan and the other leading people in his entourage. Depending on the interest groups in the court, a prediction by chief-*munajjim* or an astronomical phenomenon itself could indeed be effective in indirectly influencing decisions in the power dynamics within the court. In this respect, when the *ahkams* were assumed to hold certain fateful and moral power in themselves they could be appropriated and used to serve political ends.

In his *ahkams* the chief-*munajjim* seems to point out prevailing conditions in court politics, positioning himself within the sphere of gravity of sultan's own power. Hence the *ahkams* are expected to produce narratives recreating and enforcing the omnipotence of the Sultan and his state machinery. In these prophecies, the chief-*munajjim* seems to refrain from mentioning any specific names but only points them out through their status in general. This preference of omitting proper nouns might indicate

---

[26] Ayduz, 'Osmanlı Devleti'nde Müneccimbaşılık ve Müneccimbaşılar', p. 32.

[27] Ayduz refers to two separate lists of *munajjims*: the first list appears in *Cedavil-i Hareket-i Ta'dil el-Şams*, a table for the movements of the Sun which was prepared by an astrologer Makami Süleyman b. Mustafa b. Kemali (d. 1795) and the second list belongs to an anonymous Horoscope Journal in the Kandilli Observatory with a catalogue number of 323.

their subject status vis-a-vis the Sultan as well as the understandable necessity of using wooly expressions in order not to cause any potential turmoil in state affairs, should the predictions emphasize any negativity or enmity.

Chief-*munajjim* Fethiyeli Halil Efendi

Being an author of 23 annual predictive works, Halil Efendi lived his youth during the relatively peaceful Tulip Period (1718-1730) when Ottoman culture began to orient itself towards the Western world. Having completed his madrasa education along with studying astronomy (*Ilm-i heyet*) and astrology (*Ilm-i nudjum*), he became first a *mudarris* and later *munajjim-i sani* (vice astronomer/astrologer) and finally *müneccimbaşı* (chief astronomer/astrologer) in the court in 1746. [28]

We can presume a close relationship between chief-*munajjim* Fethiyeli Halil Efendi and Mustafa III, especially within the context of the war with the Russians, which started in 1768. It is so striking that after war was declared in 1768, the date of the beginning of the campaign coincides with the very same day when Mercury went direct in its motion on 27 March 1769. So it may have been an elected time by the chief-*munajjim* himself. For an astrologer's reasoning, this should not be surprising since Mercury rules movement, transport and communication in astrological terms.[29]

In addition to preparing *ahkams*, annual predictive works and making calendars, *munajjims* were also expected to determine auspicious dates and times for specific events. This could be well for a circumcision ceremony, a reception of an ambassador, or for example, launching a galley as seen below:

Fethiyeli Halil Efendi writes the following note (Fig. 1):

This is a memorandum by chief-astrologer for launching a galleon
God willing, with the help and by the grace of God, launching of a three kantars [a specific weight measure] galleon has been found favorable when Jupiter is the ruler of time on January the 9th, 1758 at

---

[28] Aydüz, 'Osmanlı Devleti'nde Müneccimbaşılık ve Müneccimbaşılar', p. 145.
[29] Halil Efendi's estate, which illustrates a long list of the books in his library, contains books of astrology like Al Biruni's *Elements of Astrology* (*Tefhim*) as well books by Abu Masher (*fi'n Nücum*) and Buhari (*Eşcar ü Esmar*).

04:20 and on January the 12th, 1758 at 05:05 and this is bowed respectfully before sublime audience. [30]

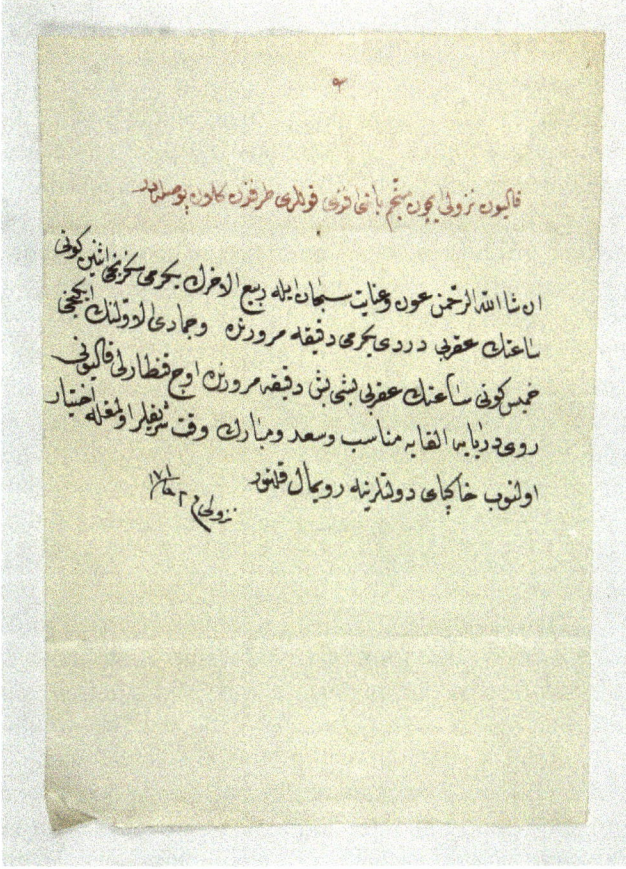

Fig. 1. A memorandum by chief-astrologer for launching a galleon.

---

[30] A.E III.MUSTAFA 3625
'Kalyon nüzûlü için Müneccimbaşı efendi kulları tarafından gelen pusuladır.
İnşaallahürrahman avn ve inâyet-i sübhân ile Rebiülahirin yirmi sekizinci isneyn
günü saatin akrebi dördü yirmi dakîka murûrunda ve Cemaziyelevvelinin ikinci
hamîs günü saatin akrebi beşi beş dakîka murûrunda üç kantarlı kalyonu rûy-ı
deryâyâ ilkâya münâsib ve sa'd-i mübârek vakt-i şerîfler olmakla ihtiyâr olunup
hâk-i pây-i devletlerine rûy-mâl kılınır.
Nüzûlü fî 2 Ca sene [1]171'.

Ahkams

Preparing and presenting *ahkams* to the Sultan and his entourage every *Newruz* was an essential part of court *munajjim*s' duties. When the structure of yearly *ahkam*s is studied, we come across a certain type of cosmological framework that gives *ahkams* their own form. Mainly two separate parts can be distinguished here:

The first part of the *ahkams*, beginning with Sultan's current conditions, gives different entries for each hierarchical layer of the court including *Reaya* in a top down manner. The usual line-up is composed of predictions for the grand vizier, other viziers and courtiers, the conditions of armies and commanders, harem, members of *Ilmiye* Class, including the chief-*munajjim* himself, and later merchants, conditions of food and their price-fixings (*narhs*), weather conditions, and circumstances of the lands and people which rest beyond the periphery of the state.

The second part is composed of monthly predictions. In this part, the chief-*munajjim* depicts certain acts as good or bad depending on the position of the Moon, which is called elections (*ikhtiyarat*). There are also astronomical phenomenon like planetary ingress into zodiacal signs and retrograde movements, and their astrological significators, in this part.

A sample of ahkams

When studying Halil Efendi's *ahkams*, I intentionally tried to compare his writings with the events that took place before he wrote down the *ahkams* and usually found that he had pursued a way of prediction depending on earlier events that had taken place. As expected, he was interpreting astrological symbols in parallel within this pre-set atmosphere and the prevailing court politics. For example, in one of his notes relating to the Jupiter-Saturn conjunction in 1762, at the time when Koca Ragıp Paşa was still alive, he put forward astrological testimonies against any military expedition, which seems to be in line with Koca Ragıp Paşa's known attitude. The time frame of this astrological judgement coincides strikingly with both the Seven Years' War and the later years of the tenure of Mustafa III's favourite Grand Vizier, Koca Ragıp Mehmed Pasha (1698-1763), who had served the Empire for decades as an administrator and diplomat in war and peace. According to the chronicler Şem'dani-zâde Fındıklılı Suleyman Efendi, Ragıp Paşa Pasha was also a close mentor to the sultan and prevented him from rushing into unwise decisions.[31]

---

[31] M. Munir Aktepe, *Şem'dani-zâde Fındıklılı Süleyman Efendi Tarihi, Mür'i't-Tevarih*, II (Istanbul Universitesi Edebiyat Fakültesi, 1978), p. 4.3

Henning Sievert, referring to *Mür èi't-Tevarih*, indicates that the sultan had an impulsive character and was eager to participate in the Seven Years' War.[32] In this respect, the crucial aim of this additional judgement on the conjunction by the *munajjim* might have an undeniable effect on the sultan's decision regarding entry into the war. On the *munajjim*'s side, while his astrological judgements first imply a potential 'yes' answer for a military campaign, he finally concludes in favour of a stay (*ikamet*), supporting his reading through an earlier interpretation expressed in Arabic.[33] His first remarks on the potential for a campaign could be seen as a justification of the Sultan's eagerness for a war; however, he immediately asserts that such a campaign has already been blocked for astrological reasons.[34] In fact, the chief-astrologer's attitude seems to be finely aligned with the position of the Grand Vizier, whose mentorship over the sultan was evident. Halil Efendi, at least from his written works, seems to have played his role strictly within the confines of court politics.

It is also striking to note that when Fethiyeli Halil Efendi's *ahkams* are listed in terms of the years in which they were produced, we cannot find the *ahkam* relating to the year 1768 when war was declared on the Russians. For reasons unknown to us, the *ahkam* of 1768 was not included in the manuscript collection in the Topkapı Palace. We might never know why this crucial year's *ahkam* was omitted, but speculatively, perhaps, it may have been removed and destroyed perhaps by Fethiyeli Halil Efendi himself, because of the prevalent discussions prior to the war.

Furthermore, it's very noticeable that the *ahkam* of 1769 does not include a specific reference for Shaik'ul Islam, which in fact seems unusual in its form. Instead the text advances with general titles of *ulema*, *mudarrises*, judges and *seyyids*. In terms of the political dynamics in the Ottoman court prior to war against the Russians, Uzunçarşılı writes that Shaik'ul Islam Veliyyüddin Efendi was on the side of Grand Vizier Muhsin-zade Mehmed Paşa on the perils of waging war promptly without

---

[32] Henning Sievert, 'Favouritism at the Ottoman Court in the Eighteenth Century', in *Court Cultures in the Muslim World, Seventh to Nineteenth Centuries*, ed. Albrecht Fuess and Jan-Peter Hartung (Oxford: Routledge, 2011), p. 282.

[33] Halil Efendi writes that the 10th house ruler (which denotes the sultan himself) is the current time-lord, which in this case is denoted by the Moon aspecting Jupiter in square aspect, hence indicating a campaign. Astrologically the Moon usually signifies voyages and movement. However, as Saturn is conjunct with Jupiter in his own 10th house, it shows a hindrance.

[34] 'Yet it hinders campaigns as it located near to the Saturn in the tenth house', '*Lâkin beyt-i aşirlerinde Zuhal'e mukârin olmakla sefere mânié olur*'.

fortification of fortresses and necessary arrangements.[35] Given the fact that Mustafa III's pro-war attitude, the death of Veliyyüddin Efendi and the assignment of Piri-zade Sahip Molla, who had similar ideas on war as Sultan Mustafa III, and the dismissal of Grand Vizier Muhsin-zade Mehmed Paşa should have shifted the gravity of political centre towards a pro-war attitude in the court.

We are not able to find what Fethiyeli Halil Efendi wrote in relation to prevailing pre-war conditions in his dislocated *ahkam* for the year 1768; however, Mustafa III's pro-war attitude must have been supported by his *munajjim's* interpretations. Later, the chronicler Şemdanizade Süleyman Efendi, writing of the accounts of losses in the war against the Russians, especially emphasizes Sultan Mustafa III's great interest in astrology with regret.[36] İsmail Hakkı Uzunçarşılı notes that Mustafa III relied on his chief-*munajjim's* interpretations in his decision to wage war against the Russians. [37]

Instead, for the time period before the campaign started in the spring of 1769, we find Fethiyeli Halil Efendi's encouraging words for the Grand Vizier's conditions, albeit with some precautionary remarks and concerns about negligent and untrustworthy *ayan*.

Fethiyeli Halil Efendi describes the Grand Vizier's conditions for the Spring time as follows (Fig. 2):

Conditions of the Grand Vizier and chief-commander and of other viziers and of cautious and trustworthy courtesans

---

[35] Ismail Hakkı Uzunçarşılı, *Osmanlı Tarihi*, Vol. IV (Ankara: Türk Tarih Kurumu, 2007), p. 367.

[36] Şemdanizade expresses his concerns as follows: 'and due to our sultan's and his entourage's inclination in astrology and all his decisions with astrological election, although this fleet would normally be expected to be launched in the morning hours, it was launched in the afternoon hours. People who do not trust in God, but give respect in this science would be expected to be lowly and eventually would lead to such calamities'. '*Ve sebeb-i diğer Padişahımızın sevk-i kurena ile fenn-i nücuma ragıp olup, sa'at-i sa'd ihtiyarsız bir işe meşgul olmadığından, iş-bu donanma Tersane'den feyz-i sabahda çıkmak adet-iken, matlubu olan sa'at vakt-i asra karib olmağla kabl-i asrda ihraç olunmuş idi. Tevekkül terk edüp, bu fenne itibar edenler süfli olup, böyle beliyyeye düçar olmaktan hali olmazlar'*. Fındıklılı Süleyman Efendi Şem'dâni-zâde, *Mür'i't-Tevârih*, vols. I-III, trans. Münir Aktepe (Istanbul: Edebiyat Fakültesi Matbaası, 1976).

[37] Şem'dâni-zâde, *Mür'i't-Tevârih*, p. 343.

Due to the Venus (*Sad-ı asgar*) in her own sign, and being the lord of farmers and lord of Reaya, and the recent conjunction of Jupiter and Saturn (*kırân-ı ulviyyîn*) happens in the 11th house, and the lord of the 11th house aspecting Jupiter (*Müşteri*), respectable Grand Vizier's requests and promises will be fulfilled by the commanders, and because of abundant ammunition, shining skills of soldiers, conquest and victory, and utmost efforts of reaya, and equipments of troops, and tenacity of vineyard owners and gripping of castles, it would increase sultan's treasure and also indicates recovery of the castles by the transfer of deputies by land and sea and removal some of loose, negligent, greedy and lier famous people and rich people (*ayan*) and compliments and good reputations to the old and faithful. [38]

---

[38]   '*Hal-i Vezir-i a'zam-ı azimü'ş-şan ve sipehderan-ı nusret unvan ve sa'ir vüzeren ve müdebbiran alallahi te'ala iktidarahüm ila yevmi'l-mizan.*
Tâli'-i fasl-ı baharda sâhib-i beyt-i hâdî aşar Sa'd-ı asğar kendi beytinde kaviyyü'l-hâl ve Sâhib-i Sehmi'l-fellâh ve Sahib-i Sehm-i hâli'r-ra'iyye bulunup ve intihâ-i tâli' sâl-i kırân-ı ulviyyîn beyt-i hâdî aşar da vukû'u ve Sahîb-i beyt-i hâdî aşar Müşterî'ye mukâbeleten nâzır ve Müşterî râci' delâlet eder. Allâhü te'âlâ a'lem. Vekîl-i muhtâr ve müşîr-i büzürgvâr hazretlerinin inân-ı irâdet ve ihtiyârları icrâ-yı evâmir-i aliyye-i Sultânî ve zimâm-ı itâ'at ve ınkıyâdları tedbîr-i mühimmât ve müşkilât-ı devlet-i pâdişâhiye masrûf olmağın mezîd takarrub ve ihtisâslarından envâr-ı suver sa'âdât-ı âyîne-i murâdâtlarında nümâyân ola ve tedbîr-i feth ve zafer ve hirâset-i ra'iyyet ve kişver ile kemâl-ı takayyüdlerine ve serdâr-ı nusret mu'tâdın techîz-i ecnâd ve ta'ciz-i ehl-i bağy ve inâd ve teshîr-i kılâ' ve bilâd ve ihrâc-ı donanma-yı hümâyûn zafer rehber ve teksîr-i hazâyin-i padîşâh-ı adâlet küster husûsunda sarf-ı kudretleri zâhir ola ve vükelâ-yı devletten ba'zıları berran ve bahren mahâfazat-ı memleket ve ebrâr ve tahlîs-i kılâ' ve diyâr içün nakl ve hareketlerine ve ba'zı diyâr ve emsârda tekâsül ve ihmâl ve sû-i tedbîr ve nâ-sezâ ahvâl ve nehmet ve tezvîr sebebiyle meşâhir ve a'yândan ba'zıların azl ve nefy ve zevâllerine ve ashâb-ı sadr ve kadrden ba'zı kâr güzâr ve emekdârlara taraf-ı şehriyâr-ı cihândârîden iltifât ve i'tibâr olunup ref'-i mazarrat-ı ehl-i fesâd ve eşrâr ve def'-i zulm ve bid'at-i ra'iyyet ve ebrâr içün istihdâm olunmalarına Allâhü te'âlâ a'lem'.

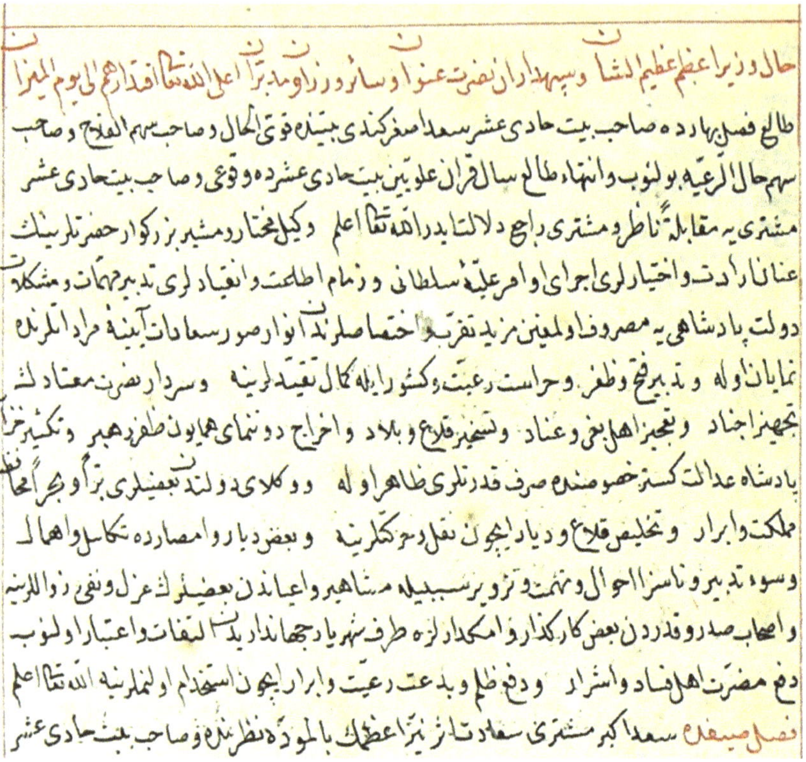

Fig. 2. Conditions of the Grand Vizier and chief-commander and other viziers.

Indeed, low morale and unprepared military forces are also highlighted in Fethiyeli Halil Efendi's *ahkams* in the separate section for the condition of armies (Fig. 3). *Munajjim* writes:

> Conditions of the soldiers and other famous commanders and lords of the districts and persons of dignity and governors

In the vernal equinox chart, Mars and Venus have been found the lords of the 11th house. And the part of lands and the part of the condition of the folks are in the 11th house and the 11th house is happily aligned with the lesser benefic, give proof the following that, *Allahu te'ala a'lem.* [Almighty lord of the worlds] famous soldiers and victorious commanders will move out promptly in alliance and arrive to those places by imperial order in order to protect and safegourd of people and to defeat the looters

and the enemy soldiers, some of them who feels worn out will be flattered
by various treatment of the sultan and their future stars of power will be
intact but who are those lousy and shabby and ignoble will be away from
merciful regard and will be lowered in rank...[39]

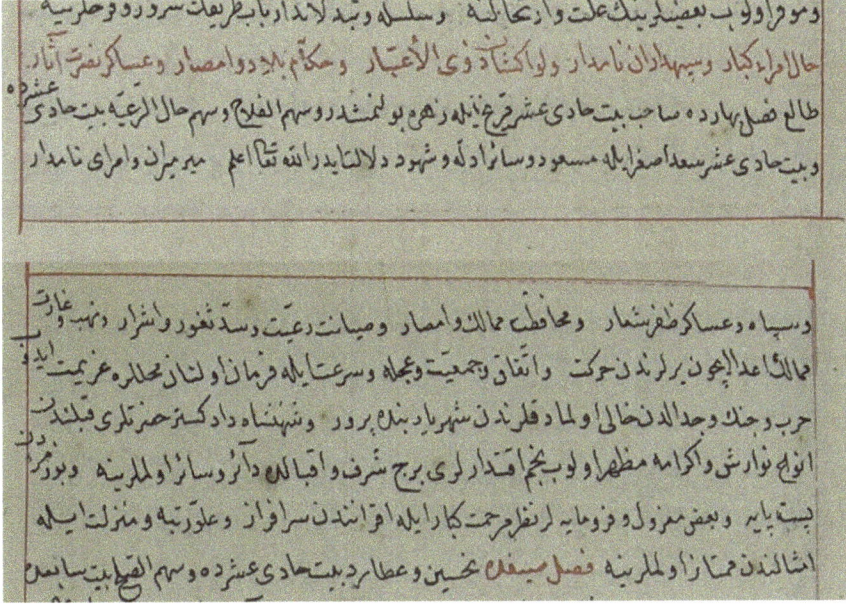

Fig. 3. Conditions of the soldiers and other famous commanders.

---

[39] 'Hal-i umera-i kiram ve sipehderan-ı namdar ve liva Keşan ve zevi'l-itibar ve
hukkam-ı bilad ve emsar ve asakir-i nusret asar.
Tali-i fasl-ı baharda sahib-i beyt-i hadi aşar Merih ile Zühre bulunmuşdur. Ve
Sehmü'l-fellah ve Sehmü hali'r-ra'iyye beyt-i hadi aşarda ve beyt-i hadi aşar
Sa'd-ı asgar ile mes'ud ve sa'ir edille ve şuhud delalet eder. Allahü te'ala a'lem.
Mir-i miran-ı ve umera-yı namdar ve sipah ve asakir-i zafer şi'ar ve muhafazat-ı
memalik ve sıyanet-i ra'iyyeti ve sedd-i sügur ve eşrar ve nehbb ü garet memalik-i
a'da içün yerlerinden hareket ve ittifak ve cem'iyyet eve acele ve sür'at ile ferman-
ı olunan mahallere azimet edip harb ve cenk ve cidalden hali olmadıklarından
şehriyar-ı bende perver ve şehenşah-dad küster hazretleri kıbelinden enva'ı nevziş
ve ikrama ve mazhar olup necm-i iktidarları burc-ı şeref ve ikbalde da'ir ve sa'ir
olmalarına ve zümreden pest-paye ve ba'zı ma'zul ve furumayeler nazar-ı
merhamet kibar ile akranından ser efraz ve uluvv-i rütbe ve menzilet ile
emsalinden mümtaz olmalarına'.

Through these explanations, we can conclude that the *munajjim* was well aware of the conditions and prevailing political atmosphere in the court. His remarks on the poor conditions of the soldiers and his following prediction that the Sultan would soon reward them can be read as an implicit political hint for the head of military forces. In a sense that the *munajjim* seems to be concerned about ongoing debates around the weak state of the military forces, however he also tries to appear not to fail to motivate them in line with the Sultan's own wishes.

The *Munajjim*'s advisory role in the court culture

Obviously, *munajjims* as a part of the state bureaucracy undertake a critical position which voices the hopes and expectations of court politics. Depending on the closeness to the Sultan and his personal interests, we could say that from time to time, the *munajjims* would play the role of a *musahib* (companion) who could influence the ruler's decisions, which in turn could create arbitrariness in patrimonial structures.[40]

Often we find them expressing soothing words and taking a deliberate cautioning approach when their astrological findings contain dangerous significations. As for Fethiyeli Halil Efendi, taking into account his long tenure, we could assume that his style in writing his *ahkams* should be expected to be controlled, cautious and diplomatic, and mostly in nebulous words in order not to cause any potential turmoil in state affairs should predictions emphasize any negativity or enmity. In his advisory role, and probably in his close relation to Mustafa III, he is seen as divinely legitimizing the Sultan's decisions, as in the decision to wage war against Russia; a contrary case should be unthinkable, indeed.

Bearing in mind that the introduction of modern astronomy, which is based on premises of a heliocentric universe model, was not entirely accepted and adapted by Ottoman astronomers until the mid-eighteenth century, the *munajjim*'s advisory role was still intact and, as in the special case of Mustafa III's interest, Fethiyeli Halil Efendi, being a member of Ilmiye class, was an active courtier in his entourage. Indeed widespread belief in astrological explanations could be found among the ruling elite. For example, Ahmed Resmi Efendi, Mustafa III's ambassador to Frederick the Great of Prussia mentions the conjunction of Mars and Saturn in the

---

[40] Halil İnalcık, *Decision making in the Ottoman State,* Essays in Ottoman History (Istanbul: Eren, 1998), p. 118 .

sign of Cancer as one of the reasons of the debilitating war against the Russians in his *Hulasatü'l – İtibar.* [41]

Finally, in parallel to the expansion of the Ottoman *Ilmiye* Class in the eighteenth century, Harun Küçük argues that an early Enlightenment flourished in the Ottoman court during the time of Ahmet III, indicating that Naima's writings were in fact built on Naturalism, which had contained astrological reasoning based on an Aristotetalian world view.[42] For him, the practice of history and *Ilm-i nudjum* were deeply intertwined so that rebalancing the four humors of the Ottoman body politics was a necessity against corruption and decay. By readdressing earlier Muslim statecraft of the Abbasid Dynasty and their close interaction with astrologers in the court, he seems to imply that an astrological explanation of the workings of state-craft is inherently upheld in the concept of a circle of justice which foresees equity within a circle of social, economical and political elements reminiscent of an Aristotelian political ideal.

Within this larger perspective, the tradition of the practice of *Ilm-i nudjum* and its established nature was a natural product of court culture in which the *munajjims* took advisory roles. Furthermore, their *ahkams* can be seen as chronicles of future hopes and self-validating, self-legitimizing tools of state politics.

---

[41] Dr. Osman Köksal, ed., *Ahmet Resmi Efendi, Hulasatü'l İtibar* (Istanbul: Gazi Kitabevi, 2011), p. 53.
[42] Harun Küçük, *Early Enlightenment in Istanbul* (PhD thesis, University of California, San Diego, 2012), pp. 65-66.

Trystes Cosmologiques:
When Lévi-Strauss Met the Astrologers

Graham Douglas

Abstract: In October 1969 the famous anthropologist Claude Lévi-Strauss gave an interview to the well-known French astrologers André Barbault and Dr Jean-Paul Nicola for the astrology magazine *L'Astrologue*. To the author's knowledge this interview has never been discussed in academic journals, and is here published for the first time in English translation. It is considered in the context of its time, and of the issues discussed: the Surrealist movement, which had an important influence on Lévi-Strauss's early work; the structure of the unconscious mind; and the question of causation in astrology.

At the end of the interview Lévi-Strauss suggested a joint project with his interviewers to study the interpretations of serious astrologers as a way of understanding how their minds work. According to Dr Nicola, the suggestion was never developed because in his opinion there was no chance of getting astrologers to agree on how to go about it. In the last 20 years however, several theses have been devoted to similar projects.

In 1969 the anthropologist Claude Lévi-Strauss gave an interview to the French astrology magazine *L'Astrologue*, a year after talking to the surrealist magazine *L'Archibras*.[1] While the latter interview is little-known among academic researchers today, the former seems to have disappeared below the horizon – Emmanuelle Loyer who recently published a 900-page biography of Lévi-Strauss makes no mention of it, and Philippe Descola, one of Lévi-Strauss's students and now professor at the Collège de France,

---

I would like to acknowledge the help of Trudie Charles at the Astrological Association Library for supplying me with scans of articles which are hard to locate elsewhere.

[1] The original interview in French appeared in the journal *L'Astrologue* 9 (Autumn 1970): pages 1–6. I am grateful to André Barbault for permission to publish my translation, and to both interviewers for their replies to my queries. Claude Lévi-Strauss, 'Voix Off, Claude Courtot entretien avec Claude Lévi-Strauss', *L'Archibras* 3 (1968).

Graham Douglas, 'Trystes Cosmologiques: When Lévi-Strauss met the astrologers', *Culture and Cosmos,* Vol. 18, no. 2, Spring/Summer 2014, pp. 143-166.
www.CultureAndCosmos.org

expressed surprise to me that he would have done such a thing.[2] Both interviews are indexed on the site http://las.ehess.fr/index.php?1337, which shows that Lévi-Strauss was rather a prolific subject.

The interview was nearly 50 years ago, but the interviewers have been able to clarify some points regarding the context. In a letter to me, which he dated by the position of the sun in the zodiac, André Barbault said that it was his initiative to write to the famous anthropologist, admitting that he was surprised to receive a reply. And Dr Jean-Pierre Nicola told me that the published interview was considerably shortened by editing out his argument with Lévi-Strauss over causality. He says that the latter did eventually accept that there might be causal effects operating.[3]

At the time that he gave these interviews Lévi-Strauss was at the peak of his fame and even notoriety, according to his biographer: *Playboy* magazine ran an article about structuralism – which Lévi-Strauss responded to – and the trainer of a football team said he wanted to use structuralist principles.[4]

Since Lévi-Strauss was advised by the astronomer Jean-Claude Pecker, a virulent opponent of all things astrological, it would seem unlikely that he would have an interest in contemporary astrology, but as Descola pointed out, Lévi-Strauss was always a man with an unusual degree of curiosity and open-mindedness.[5]

And we can see that Lévi-Strauss would be open to astrology in general terms from his remarks in *Tristes Tropiques*:

> We gave up sun worship a long time ago and we have lost the habit of
> associating the points of the compass with magic qualities, colours, and
> virtues. But... we cannot prevent the major astronomical and meteorological

---

[2] Emmanuelle Loyer, *Lévi-Strauss* (Paris: Flammarion, 2015); Emmanuelle Loyer, personal communication, October 2015; Descola email to Graham Douglas July 2015.

[3] Letter to Graham Douglas dated 1CN15 meaning 23 June 2015.

[4] Loyer, *Lévi-Strauss*, p. 577.

[5] Descola email to Graham Douglas July 2015; Pecker dismissed the careful statistical analysis by the Gauquelins as 'biased and of no value', *Bulletin de la Société Astronomique de la France*, July 1974, cited by Barbault in *The Value of Astrology: from ancient knowledge to today's reality*, English translation (London: The Astrological Association, 2014), p. 91. For a detailed account of the controversy over the Gauquelin work, see Suitbert Ertel and Ken Irving, *The Tenacious Mars Effect*. (London: Urania Trust, 1996). And for the most up-to-date survey of research into astrology, see Geoffrey Dean, Arthur Mather, David Nias and Rudolf Smit, *Tests of Astrology* (Amsterdam: Astrologie in Onderzoek, 2016).

phenomena conferring almost imperceptible but ineradicable properties on certain areas.'[6]

And:

But are they really superstitions? I see these preferences rather as denoting a kind of wisdom which savage races practiced spontaneously and the rejection of which, by the modern world, is the real madness.[7]

Although Lévi-Strauss was still being widely celebrated as one of France's premier intellectuals in 1969, classical structuralism was coming under sustained attack from many quarters. Post-structuralism had begun in 1967 with the publication of Derrida's first book, and in 1968 a collection of critical articles was published under the title *Qu'est-ce-que le Structuralisme?*.[8] Over the next five years these attacks became quite virulent, and Lévi-Strauss was considered to have been supplanted from his position as France's foremost intellectual by Foucault and Lacan. Between the two interviews he had suffered serious health problems.[9]

Of his two interviewers, André Barbault and Jean-Pierre Nicola, the former is better known, through his books and from editing the journal *L'Astrologue* for many years. He also interviewed quite a number of well-known figures about their views on astrology for his journal.[10] Two of the interviewees, Guy Michaud and Gilbert Durand, were well-known

---

[6] Lévi-Strauss, *Tristes Tropiques* (Harmondsworth: Penguin Modern Classics, 2011), p. 122.

[7] Lévi-Strauss, *Tristes Tropiques*, p. 123.

[8] Jacques Derrida, *De la Grammatologie* (Paris: Les Éditions de Minuit, 1967); Oswald Ducrot, Tzvetan Todorov, Dan Sperber, Moustafa Safouan, and François Wahl, *Qu'est-ce-que le Structuralisme?* (Paris: Seuil, 1968).

[9] Cited by Patrick Wilcken, *Claude Lévi-Strauss: The Poet in the Laboratory* (London: Bloomsbury, 2010), p. 299.

[10] Henry Miller, in *L'Astrologue* 16 (Winter 1971): pp. 193–5; André Bréton in *L'Astrologue* (1978), reproduced in English with an introduction by Nicholas Campion in *Culture and Cosmos* 6, no. 2 (Autumn/Winter, 2002): pp. 45–56; C. G. Jung May 1954, published in *L'Astrologue* 8 (Winter 1969): pp. 193–6; the literature professor Guy Michaud in *L'Astrologue* 21 (Spring 1973): pp. 1–5), who had written several books about surrealism; plus short responses from Jean Cocteau in *L'Astrologue* 12 (Winter 1970): pp. 1–2); Werner Heisenberg, a refusal in *L'Astrologue* 20 (Winter 1972): pp. 1–2. Another literature professor, Gilbert Durand, wrote an article *Astrologie et son Langage*, in *L'Astrologue* 45/46 (Spring 1979): pp. 44–55.

academics at French universities who nevertheless were pleased to talk openly about their beliefs in or practice of astrology. Other reports surface in the Internet of interest in astrology by French academics. Lucien Malavard, a science professor at the Sorbonne, mentioning primitive classification, is reported as saying that astrology was the original social science before the name existed.[11] Nicola also has authored a number of books, obtained a PhD at the University of Hawaii (2004), and founded his own school of *Astrologie Conditionnaliste*.[12] In his journal *Astrologique* he interviewed an astronomer, Hubert Reeves, who at least considered astrology to have a worthy place in the history of science, although as Nicola pointed out to me this was before Reeves had developed his career.[13]

Nicola took a lesser part in the interview, and his work has not been translated into English, although a number of articles are available on his website. But the two men have broken off all collaboration and, according to Patrice Guinard, their continuing disputes are legendary in French astrological circles.[14]

During his interview with Barbault, Lévi-Strauss mentions that he was working on a study which showed that a series of myths with their varying structures can be related to astronomical phenomena. This corresponds to the work he published in his next book *The Naked Man*, the final volume of his series *Mythologiques*, where there is astronomical data relating to the constellations of the Pleiades and Orion.

---

[11] Jack Mandon, 'L'Astrologue: Une Femme...', *Agora* (15 April 2008): '*Lucien Malavard, prof. de sciences à la Sorbonne*: « *Je pense que les anciens ont fait en quelque sorte des sciences humaines avant la lettre par l'intermédiaire de l'astrologie. Ils ont ainsi bâti une classification des êtres, une manière d'y voir plus clair dans les comportements humains. Pour ma part, je serais tenté de situer l'astrologie à côté des sciences humaines, un peu plus loin... »*, available at http://www.agoravox.fr/tribune-libre/article/l-astrologie-une-femme-38659 [accessed 20 June 2016).

[12] Nicola's PhD is on 'Globalite en psychologie analytique et astrologie conditionaliste' (University of Hawaii, 2004). His many publications can be consulted at http://www.astroariana.com/_Jean-Pierre-Nicola_.html [accessed 10 June 2016].

[13] See *Astrologique* 4 (August 1976): pp. 1–5. Email from Pierre Nicola, May 2016.

[14] See Guinard's CURA website, which is also a mine of information for serious researchers into all aspects of astrology, at http://cura.free.fr/docum/10astrof.html, entry no. 21 for Nicola.

I will first present the interview and then discuss some of the issues which arise and consider the interview in a contemporary context.

Interview with Claude Lévi-Strauss
Professor at the Collège de France. Conducted in Paris, 6 October 1969, with André Barbault and Jean-Pierre Nicola. Translated from the French by Graham Douglas.

C. L-S.: First I will tell you a little story which goes back a long way. On the day I received my *agregation* to teach Philosophy – I received the news one morning – that same afternoon I wanted to do something which in my eyes symbolized my liberation from those long years of study. So I went to the Chacornac bookshop beside the Seine, and bought a book about astrology.[15] Not at all because I believed in it or wanted to dedicate myself to studying it, but because in my opinion it was the symbol of a great intellectual construction which was worth quite as much as those whose study I had been obliged to concentrate on for years and years. I still have it. I don't think that [buying it] was anything very important, but it has been for me a kind of magical gesture, if you like.

A. B.: Do you have a settled opinion against astrology? Are you constrained by the fact that it is a creation of the [human] mind at the stage of magical thinking, if you follow me?
C. L-S.: Yes, considering that the magical mentality appeals to all sorts of mechanisms and schemes which are true of any kind of thinking. An opinion against... frankly, if you asked me whether I believe in it, I would either say 'no', or rather give you a more nuanced opinion. Not long ago it happened that one of your colleagues sent me my horoscope, – and I believe it was someone serious, because he did not give in to the temptation to make any kind of predictions for the future. It contained a whole collection of reflections relevant to my life and my character; but all told, there was nothing that a mind with a little sensitivity and intelligence could not have taken from reading my books. So, if I had to reconstruct what he was able to do in such a case, I suppose that he began from some rather vague impressions, because he had read some of my books. Not being a specialist in anthropology, he did not enter into the details – he had felt all that from outside – and then, in order to organise it he resorted to a scheme, a sort of canvas, and since this canvas is an extremely precise structure which obeys certain rules typical of all 'canvases' of this kind, that helped him to crystallize some things which he had perceived in a rather confused fashion. Then, from this base, if you wish,

---

[15] The Chacornac bookshop was also the home of Barbault's journal *L'Astrologue* for many years.

in presenting things in this way, yes, that seems to me to be completely understandable and normal. If, on the other hand it is a question of predicting what will happen in the life of an individual, well then I would have a lot of reservations to make…

A. B.: This schema, which is an astral structure, in your eyes does it have a meaning of its own?
C. L-S.: You are asking a question which is very difficult to answer. Any scheme of organisation, whatever it might be, has an intrinsic value. I'm not convinced – I don't know any more about it – that this kind has any value unique to itself: or one which is any greater than the lines on the hand, or the [Tarot] cards. In all these cases, it seems to me that it is a question of schema which impose very rigorous constraints on the interpreter; even if they are arbitrary constraints, they force [the interpreter] to think, and to make an intellectual effort, which for them is productive.

A. B.: For our part, we go further, because we conceive an objective relation between the astral data and the psychological universe of the individual.
C. L-S.: Your position is very understandable. But for an outsider it raises fundamental problems. The first is that it is hard to understand how; it seems to me that for that hypothesis to be plausible it would have to be able to produce a certain number of intermediate mechanisms between the stars and man. The second [problem], is that in the same place at the same time, approximately, there is not just one person born, and what is true for one cannot be true for the other.

J-P. N.: But on the one hand, the common background to these beings that are similarly 'starred' [have similar astrological charts], derives from an analogical structure, which does not necessarily lead to identical outcomes; and on the other hand the astronomical canvas implies some properties – distances, speeds, rhythms… – which are in themselves, structured, structural and structuring. This organisation of the cosmos orders the values of an individual more than it organises the strategy of the interpreter. These are real structures!
C. L-S. What you call real structures, are the structures of your mind, which come into play at the time that you make your studies, and the structures of your mind are also those of all minds. I am not surprised that there can be and there are coincidences (…). Astrology is an extremely ancient science – the first – and the persistent influence that it exercises on the human mind comes precisely from what it [the mind] became interested in very early on, in the truly interesting things, in other words in correspondences.[16] Only, you see,

---

[16] In the French, the original 'rapports' could also be translated as 'relations'; it is impossible to decide, but 'correspondences' is a term used by Lévi-Strauss in

from the moment when one makes the effort to establish systems of relationships intrinsic to some domains or some worlds, you establish intrinsic systems of relationships to the astronomical universe, intrinsic to the human universe, and so on. Because it is the same mind which has worked in one framework and in the other, it is unavoidable that it perceives some homologies between these systems of relationships. I am very ready to accept that. The real difficulty begins if one attempts to establish a causal relation, if I can put it like this, between one level of relationships and another, and personally, I don't believe that it is necessary in any case. I don't feel the need to do that.

A. B.: It is what we experience in the context of the 'symbolist' school of astrology, despite the probability of a physical substrate to an astral influence. It seems normal to us that the earliest societies were naturally in tune with cyclic processes in correspondence with astronomical phenomena.[17]

C. L-S.: A correspondence undeniably: because they were people who were extremely attentive to certain types of phenomena, and because they were attentive to them, of course that played a role in their lives. I am working at the moment on unravelling an entire mythology, and showing that the relations between different myths can be explained by the astronomical references of each one to a whole series of phenomena, which between themselves are in a relationship based on transformation. They are not like this objectively; they are like this for the human mind which perceives them. So, the Moon Spots and the Sunspots are in correspondence, because they are dark marks within a bright perimeter, while the conjunction of two stars with the sun at its Heliacal Rising, these are bright marks outside of the perimeter of a dark background; and then one goes on to perihélions, and peri-sélènes, anti-hélions, and a whole list of things of this kind, in a marvellous series. Then you can show that the place of each myth in relation to any other can be determined through these astronomical references, which – they too – can be put into an objective relationship, inasmuch as they offer some intrinsic characteristics which put them into correlation or opposition one with another. That does not mean of course, that these astronomical phenomena have exercised a causal kind of influence over the minds which constructed the myths, much less on their heroes [of the myths]. They have simply constructed a system, and this system, from the point when it was constructed reveals itself as working and productive.

---

describing primitive classification, while astrologers use it to describe the correspondence of microcosm with macrocosm.

[17] The 'symbolist' school is the one developed by Barbault, but the term is not used outside France, elsewhere his approach would be recognised as typical of 'psychological astrology'.

At bottom, all systems are working and productive, because man can only think with systems. Astrology has been a great system, because it has helped man to think for thousands of years.

J-P. N.: Can it still help humanity to think?

C. L-S.: Perhaps, still, yes to a certain extent. I don't doubt – not in the same sense which you have in mind – for me, because it is a subject of study, and astrology as a subject of study can greatly enlarge our vision of man and his mental mechanisms.

J-P. N.: In any case, there is no absolute contradiction for us between a symbolic expression and physical influence.

C. L-S.: You are not obliged, it seems to me, to immediately think of a causal relationship. You could very well imagine that the astronomical configurations on the one hand, and let's say the psychological configurations on the other hand – simplifying enormously – arise from something else which determines both of them together.

A. B.: We are no less obliged in this to rely on physical phenomena which influence us, such as the phases of the Moon and the tides in the oceans.

C. L-S.: It is interesting that you mention tides, because in the end what we call circadian rhythms are rhythms which are partly biological, (and we can consequently consider that they have an incontestable psychological role), and partly physical. At the same time, you can conceive that they have a relationship which we don't understand well, or that we don't understand at all, with the rhythms of much longer periodicity which are those of cosmic phenomena; in that way you would have a means of getting round the chasm that separates the stars from the destinies of individuals.

J-P. N.: Yes, through rhythms.

A.B.: The fact is we don't know why. This is what made André Bréton say that astrology holds one of the highest secrets of the world.

C. L-S.: I knew André Bréton well – we were very close for a period of time; but I won't go as far as him. I wouldn't say that it holds [secrets] – but it is perhaps one of the signs that secrets exist which we don't understand, and I feel impelled to say, that we will doubtless never understand.

You know that there is a great piece of work that we can do together, because astrology would seem to be a subject of study, if not a study [in itself], in any case one of the most fascinating subjects of study. In order to know how astrologers think, how they reason, how they analyse, we [would] use some treatises and manuals of the kind that I went to buy a long time ago. But there would be a whole study to be made that would seem to me to be much more fascinating. Starting from concrete documents, making use of the considerable quantity of horoscopes skilfully interpreted by high quality astrologers – with the aim of attempting to reconstruct the way in which their minds work, how they engage with it. This is an ethnological task, for a psychologist who doubles as a historian of science; but still it would be necessary to know a lot of astronomy as well.

Lévi-Strauss: the turn to anthropology

Anthropology had emerged as a discipline in Britain during the 19th century, while in France it only separated itself from sociology – which Émile Durkheim had largely established with his *Rules of Sociological Method*, published in 1885 – during the interwar period.[18] It was also influenced by French surrealism during this period, which will be discussed later.

In a conversation with Didier Eribon in 1989, Lévi-Strauss again mentions his purchase of an astrology book following his *agregation*, this time attributing his action simply to a desire to celebrate his freedom and to prove to himself that he still had an independent mind.[19] In the *L'Astrologue* interview however, he describes it as a 'magical gesture', which suggests that he had by then read Marcel Mauss's book *A General Theory of Magic*, even though he was not yet set on the path to becoming an anthropologist.

In *The Savage Mind* Lévi-Strauss built on the earlier work of Durkheim and Mauss.[20] Mauss, who was Durkheim's nephew and a much more congenial personality than his uncle, was greatly admired by Lévi-Strauss, and may even have been the first influence leading him towards Social Anthropology. This seems likely since Lévi-Strauss wrote asking Mauss's advice in October 1931, shortly after his *agregation*, saying he was strongly tempted to study ethnography.[21] Thus the purchase of the astrology book happened during a crucial period in Lévi-Strauss's life.

In *Tristes Tropiques*, Lévi-Strauss says that, despite the mixture of motives propelling him towards anthropology, the 'revelation' did not happen until 1933–4 when by chance he came across the book *Primitive*

---

[18] Émile Durkheim, *The Rules of Sociological Method* (London: Simon and Schuster, 1982).
[12] Didier Eribon, *Conversations with Claude Lévi-Strauss*, trans. Paul Wissing (Chicago: University of Chicago Press, 1991), p. 10.
[20] Émile Durkheim and Marcel Mauss, *Primitive Classification*, trans. Rodney Needham (Chicago: University of Chicago Press, 1963). In this book, one of the topics they studied was Chinese astrology. The book is cited by Vilhena, *The World of Astrology*, pp. 6, 21–22.
[21] Loyer, *Lévi-Strauss*, p. 115.

Society by Robert Lowie.[22] Among his other reasons were a disenchantment with the aridity of the philosophy he was expected to teach, a need for adventure and physical contact with nature and people, and simply as an escape route from his conventional life.[23]

Although Lévi-Strauss had criticised Paul Nizan's book *Aden Arabie*, he appreciated its author's search for a more authentic way of life than what was offered by French society. It helped to crystallize his belief that nature must be appreciated as it is and not reduced to a theatre for human exploitation.[24] Earlier than this, both the arts and social sciences were appealing to Lévi-Strauss. While teaching in secondary schools he enjoyed citing Baudelaire, and as a boy had been deeply impressed with Wagner's polyphonic music, which much later he came to recognise as a source of his interest in the architecture of myth.

Primitive classification and the structure of astrology

Lévi-Strauss's work in *The Savage Mind* has been used by the Brazilian researcher Luís Rodolfo Vilhena to analyse the system of astrological symbols, whose conclusions are worth summarising.[25] Vilhena begins by pointing out that although Durkheim and Mauss had initiated the study of astrology as possessing an autonomous logic, they still viewed it as an inferior stage in the development of the human mind.[26] In contrast, as is well-known, Lévi-Strauss accorded no logical or developmental priority to modern science over primitive magic.[27]

Vilhena goes on to describe what he calls the synchronic classifications within the astrological symbol system, made up of a variety of oppositions such as Fire/Water and Earth/Air, as well as positive/negative, above/below, masculine/feminine. In addition, there is the ternary system which divides the 12 zodiac signs into Cardinal, Fixed and Mutable types, and the diachronic structuring arising from the passage of the sun from

---

[22] Lévi-Strauss, *Tristes Tropiques*, p. 59.

[23] Loyer, *Lévi-Strauss*, p. 91.

[24] Paul Nizan, *Aden Arabie* (Paris: Rieder, 1931; 2nd edition with introduction by Jean-Paul Sartre, Paris: Editions Maspero, 1960); Loyer, *Lévi-Strauss*, pp. 91, 121.

[25] Vilhena, Luis Rodolfo, *O Mundo da Astrologia: estudo antropológico* (Rio de Janeiro: Jorge Zahar Editor, 1990), English translation by Graham Douglas, *The World of Astrology: an ethnography of astrology in contemporary Brazil* (Ceredigion: Sophia Centre Press, 2014).

[26] Vilhena, *The World of Astrology*, p. 24.

[27] Lévi-Strauss, *La Pensée Sauvage* (Paris: Plon, 1962), p. 24, and see Vilhena, *The World of Astrology*, pp. 23–24.

each sign to the next during one year. Vilhena acknowledges the usefulness of Lévi-Strauss's distinction between the way that these classification systems work with signs and images derived from immediate experience, in contrast to modern science which derives abstract concepts and seeks structure at a level behind appearances.[28] Later Vilhena draws attention to the way that astrological structure is maintained, in a similar way to what Lévi-Strauss found in myth, irrespective of the content of its symbols; the Sun and Moon are always seen as a contrasting pair for example.[29] And, again in line with Lévi-Strauss's analysis of myth, he points out how the astrological system of symbols, although based on a limited number of terms, can yet produce almost infinite possibilities of interpretative combinations.[30]

Vilhena's work is the first full-length sociological analysis of the function of astrological practice in the lives of its contemporary students, and he goes on to point out how modern astrologers, in common with people in non-industrial societies, are concerned to view the human being and the events of their lives as meaningfully structured wholes, and less through the lens of causality or mechanism. He cites Lévi-Strauss remarking that in systems based on dichotomous oppositions, 'everything has a meaning, otherwise nothing has a meaning'.[31] The sociological context of the practice of astrology has been much studied since then and it will be considered later in relation to the suggestion by Lévi-Strauss in the interview of making a study of astrological interpretations.

Returning to the interview, we see that Lévi-Strauss resists the astrologers' claim that there is an intrinsic meaning in astrology, and says that in his view they have no need of a causal model, since the correspondences that they perceive arise from the structuring of their own minds. This is fully in line with the Structuralist refusal to see cultural phenomena as determined by social or biological ones.[32]

Looking at the book *The Value of Astrology*, which Barbault considers his 'testament', and has recently been made available in English, his comments on astrology and psychoanalysis accept that astrology originates within the human being, and that what we observe in the world is a

[28] Vilhena, *The World of Astrology*, pp. 27–30.
[29] Vilhena, *The World of Astrology*, p. 48.
[30] Vilhena, *The World of Astrology*, p. 52.
[31] Vilhena, *The World of Astrology*, p. 58; and Claude Lévi-Strauss, *La Pensée Sauvage* (Paris: Plon, 1962), p. 228.
[32] See http://www.cardiff.ac.uk/socsi/undergraduate/introsoc/durkheim4.html for more on this distinction which goes back to Durkheim [accessed 16 Jun 2016].

projection of that.[33] Barbault cites Jung saying that the personality, being unconscious, cannot be distinguished from the contents of its projections, 'which means that to a great extent it is identical to its surroundings'.[34] However Barbault also says that 'man, by introjection, astralises himself and assimilates himself into a celestial body', which may have been difficult for Lévi-Strauss to accept, along with the claim that the analogy between the human soul and nature and the constellations was 'how the structuralist seminology (*sic*) of astrology came about in the universe of the word'.[35] Whatever this 'introjection' means, any correspondences between the psyche and patterns projected onto the heavens do not require the movements of the planets to somehow correlate with movements in the psyche, yet this is clearly what Barbault intends since he devotes a whole chapter to astrological forecasting.[36] Lévi-Strauss states explicitly in the interview that he does not accept forecasting, nor would he go along with Barbault's attempts to employ Freudian developmental language to explain the meanings of the planets – 'Mars (the oral-sadistic stage energy, according to the Freudians' vision)'.[37]

Lévi-Strauss and Jung

Reading the interview today, it seems strange that Jung was not discussed, given how central his psychology has become to modern astrology.[38]

---

[33] André Barbault, *The Value of Astrology: from ancient knowledge to today's reality* (London: The Astrological Association, 2014), translated by Kate Johnston from *L'Astrologie Certifée* (Paris: Éditions du Seuil, 2006), Translator's preface, and pp. 163–165.

[34] Barbault, *Value*, p. 167, citing Jung, 'The Roots of Consciousness', no page or publication details given.

[35] Barbault, *Value*, p. 166. The French word 'verbe' can be translated as 'word', 'voice' or 'verb'. The meaning of this sentence is quite obscure, but it is possible that 'verb' is a better translation because Barbault was familiar with the work of Gilbert Durand, who described his work as 'structuralist' and emphasized that the archetypes were active entities like verbs in grammar. Barbault has confirmed to me that he was indeed influenced by Durand's thought in this passage, (email dated '8 degrees Gemini, 2016'). See also Mable Franzone, 'L'Imaginaire: une approche de la pensée de Gilbert Durand', trad. Marilyne Renard, *Esprit Critique* 5, no. 3 (Summer 2003): p. 7, available at
www.espritcritique.fr/0503/esp0503article05.html [accessed 2 July 2016].

[36] Barbault, *Value*, pp. 118–159.

[37] Barbault, *Value*, p. 173.

[38] Dr Nicola has confirmed that Jung was not mentioned at all during their conversation, email to Graham Douglas May 2016.

However, at the time of the interview the only work that had incorporated Jungian psychology into astrology in the English language was Dane Rudhyar's *Astrology of the Personality*, originally published in 1936. Most of the influential books on 'Jungian astrology' did not appear in English until the 1970s.

Lévi-Strauss said that Freud's psychoanalysis immediately appealed to him, by confirming his own distaste for dry logical categories that were 'no more than a gratuitous intellectual game', and that the oppositional categories of philosophy such as rational and irrational did not exist in that form. Instead he saw that meaning was a more important category, and one which was intimately emotional.[39] But Lévi-Strauss's disapproval of Jung's formulation of the archetypes of the collective unconscious, and his dislike of Jung's 'mysticism' are well-known, although he made few comments in print.[40]

Both Jung and Lévi-Strauss were deeply interested in the unconscious mind, and the resemblances between their respective structural analyses have been claimed to be more than coincidence. Thus D'Aquili quotes a long section from *Structural Anthropology*, pointing out that both the archetypes of the collective unconscious (Jung's term), and the unconscious mind (Lévi-Strauss's term) were conceived of as empty structures, which only became active when an emotionally charged situation brought their structuring influence to bear upon the personal issues in play for an individual.[41] And D'Aquili goes on to remark on the extraordinary resemblance of this passage to statements from Jung's work from twenty or more years earlier. For example, talking about certain 'inherited psychic factors', 'universal dispositions of the mind', 'which can also be described as categories analogous to the logical categories which are always and everywhere present…' he says that: 'only in the case of our "forms" we are not dealing with categories of reason, but with categories of the imagination'. 'The original structural components of the psyche are of no less surprising a uniformity than those of the physical body. The archetypes are, so to speak, organs of the pre-rational psyche. They are eternally inherited forms and ideas, which have at first *no* specific content.

---

[39] Lévi-Strauss, *Tristes*, p. 55.

[40] Lévi-Strauss, *Pensée*, p. 88.

[41] Eugène d'Aquili, 'The Influence of Jung on the work of Claude Lévi-Strauss', *Journal of the History of Ideas* 11 (1975): pp. 41–48; and Claude Lévi-Strauss, *Structural Anthropology* (New York: Anchor, 1963), pp. 198–199.

The specific content only appears in the course of the individual's life, when personal experience is taken up in precisely these forms'.[42]

D'Aquili identifies other remarkable parallels in the treatment of kinship by the two authors, noting that Lévi-Strauss published his work three years later than Jung's.[43]

Likewise, the familiar quaternary structure of four elements, which Jung attributed to the collective unconscious, was also a feature which Lévi-Strauss found in certain myths, in one Navajo example the specific contents – Fire, Earth, Wind and Water – are almost identical to those of ancient Greek cosmology. Lévi-Strauss rejected the tendency of Jung to believe in the *essential* nature of these archetypes, whereas Lévi-Strauss always emphasized that it was structural relations between terms that were the key, not the terms themselves, and further, that it was the relationship of transformation between systems that was the key feature of structuralist analysis.[44] But Jung's essentialist conception of the archetypes had shifted to something much closer to the structuralist view at least 17 years before Lévi-Strauss attacked him.[45] Lévi-Strauss's attitude persisted in an interview with George Steiner in August 1966, although according to Staude he later moderated his view, and said that Jung had 'become open to a much more formal structuralist approach'.[46]

Richard Gray has reviewed d'Aquili's article, noting that despite these parallels Jung's approach differed from that of Lévi-Strauss in accepting more than purely synchronic abstract data, and moving – as a therapist of course – into subjective areas of the patient's experience.[47]

While it is quite plausible to link the four astrological elements to Jung's four psychological functions, some Jungian astrologers have attempted to identify features in the birth chart with certain archetypes, but

---

[42] Cited in D'Aquili, p. 45, from C. G. Jung, *Psyche and Symbol* (New York: Doubleday Anchor Books, 1958), pp. 292–293. Italics by D'Aquili.

[43] D'Aquili, p. 47.

[44] Eribon, *Conversations*, pp. 162–163.

[45] D'Aquili, p. 48.

[46] Cited in J. R. Staude, 'From Depth Psychology to Depth Sociology: Freud, Jung and Lévi-Strauss', *Theory and Society* 3, no. 3 (1976): pp. 303–338, footnotes 65 and 66; George Steiner, 'A Conversation with Claude Lévi-Strauss', *Encounter* 26 (August 1966): p. 52.

[47] Richard M. Gray, 'Jung and Lévi-Strauss Revisited: An Analysis of Common Themes', *The Mankind Quarterly* XXXI, no. 3 (Spring 1991).

they disagree on how it should be done, or if it should be done at all.[48] It is also true that Jung tended to speak of archetypes and archetypal images, although the latter are are much more numerous, quite indiscriminately. In an interview in 1954 Jung said that 'the "planets" are the gods, symbols of the collective unconscious', without referring to archetypes such as the Persona or the Shadow.[49]

Another important stage in the wider structuralist trajectory – although his work was not translated into English until 1958, and French in 1965 – was Vladimir Propp's analysis of the Russian folktale.[50] Lévi-Strauss was not impressed by the application of structuralist approaches to literary analysis, and treated Roland Barthes' work almost with contempt.[51] But it can be argued that Greimas's development of Propp's work on the structure of narratives based on a group of six actants, has a more direct connection with astrology.[52] Roles such as the Hero, the Desired Object, the Helper and the Obstacle seem to have some similarity with the astrological archetypes of Mars, Venus, Jupiter and Saturn respectively, while Greimas's Sender and Receiver plausibly suggest the active Sun and the receptive Moon.[53]

Gilbert Durand proposed a scheme in which the planets and zodiac signs are grouped with various archetypal images on the basis of a concept of symbolism rooted in the main postures and actions of the human body.[54]

---

[48] See Maggie Hyde, *Jung and Astrology* (London: Aquarian Press/Thorsons, 1992).

[49] Jung responded in writing in May 1954 to a questionnaire by André Barbault and Jean Carteret, a transcript of his replies was published in Barbault's journal, *L'Astrologue* 8 (Winter 1969): pp. 193–196, and an English version is included in C. G. Jung and Gerhard Adler, *C.G. Jung Letters*, trans. R. F. C. Hull (London: Routledge, 1976), Vol. 2, pp. 175–7.

[50] V. I. Propp, *Morphology of the Folktale* (Leningrad, 1928; 2nd revised edition, Austin: University of Texas Press, 1958).

[51] Loyer, *Lévi-Strauss*, pp. 580–583.

[52] A. J. Greimas was a French-Lithuanian who developed a structuralist analysis of semantics, using what became known as the Semantic Square, derived from the Logical Square of Aristotle. For more see Terence Hawkes, *Structuralism and Semiotics* (Abingdon: Routledge, 1977), pp. 87–95.

[53] Graham Douglas, *Physics, Astrology and Semiotics* (published by the author, 1983); 'Greimas's Semiotic Square and Greek and Roman Astrology', *Semiotica* 114, nos. 1 and 2 (1997): pp. 1–19.

[54] Gilbert Durand, *Les Structures Anthropologiques de l'Imaginaire: introduction à l'archetypologie* (Paris: Dunod, 1992). See also: M. Maffesoli, ed., *La Galaxie*

Greimas rejected Durand's approach since it views symbols as determined by a base in the organism, and cannot therefore be considered a structuralist method.[55] On the other hand, Durand's approach seems quite in tune with the cognitive semantics school, in which bodily schema are seen as central to image making, and Jung acknowledged a similarity between archetypes and innate animal behaviours.[56] As is well-known, the zodiac signs have always been related to different features of the human body, and Louis Cruchet has identified close resemblances between Durand's scheme and the Conditionalist astrology proposed by Jean-Pierre Nicola.[57]

In emphasizing the dynamic nature of psychic images and archetypes, and refusing the reduction of symbols to signs, Durand's brand of structuralism – if that is what it is – seems to offer a theory of symbols closer to the experience of astrologers.[58] Durand was also the Jury Chief for Patrice Guinard's PhD thesis at the Sorbonne, and although Guinard was not influenced by him, Durand told him that in his book *La Foi du Cordonnier* he had attempted something similar.[59]

Surrealism

André Breton's name crops up only briefly in the interview, but he had an important influence on Lévi-Strauss. Lévi-Strauss had met Bréton and the Cuban surrealist painter Wifredo Lam on the boat which took them from Europe to Martinique, and credited Breton with opening his eyes to the

---

de l'Imaginaire: Derive autour de l'oeuvre de Gilbert Durand (Paris: Berg International, 1980).

[55] A. J. Greimas, *Semantique Structurale* (Paris: Larousse, 1966).

[56] For a discussion of Jung's views by one of his close associates, see Jolande Jacobi, *Complex/Archetype/Symbol in the Psychology of C.G. Jung* (Bollingen Series LVII) (Princeton: Princeton University Press, 1959), pp. 39–46.

[57] Louis Cruchet, *Ethnoastronomie at traditions astrologiques* (Paris: Publibook, 2009), pp. 161–179. He also cites Nicola as saying that human astral determinism is biological not psychological, see p. 165.

[58] See Maggie Hyde, Chapter 4, 'The Symbolic Attitude', in *Jung and Astrology* (Aquarian/Thorsons, 1992), available at:
http://www.maggiehyde.com/books/jung_sings.html [accessed 3 July 2016].

[59] Email from Patrice Guinard, 1 Jun 2016. Guinard's thesis is entitled 'L'Astrologie, fondements, logiques et perspectives' (Paris, The Sorbonne, 1993). It is available on his website http://cura.free.fr, along with a list of other theses on astrology over the last 100 years: http://cura.free.fr/01authd.html. Gilbert Durand, *La Foi du Cordonnier* (Paris: Editions Harmattan, 1984)

exotic hidden in the ordinary.[60] In fact, the surrealist attitude of looking for the marvellous within the ordinary, and creating new perspectives by juxtaposing objects or images that were normally considered incongruent goes to the heart of the developing methods of anthropology in France during the 20s and 30s.[61] Surrealists too looked for inspiration among marginal areas of society, and Lévi-Strauss described the anthropologist as a *Chiffonnier* or rag-and-bone-collector. His famous use of the term *bricolage* in *The Savage Mind*, to describe the ways of thinking in non-industrial societies, refers to the use of whatever comes to hand as a means of creating artefacts.[62] The crossover between surrealism and ethnology accelerated among the émigrés in New York, but it began well before in Paris, as evidenced by the contributions of the surrealist Michel Leiris and the anthropologist Marcel Griaule to the short-lived arts magazine *Documents* in 1929–30. And Leiris set off in 1931 as the secretary on Griaule's 2-year Dakar-Djibouti expedition, which led to huge interest among the French public when it returned and its artefacts and recordings were displayed in an exhibition.

Leiris later became an anthropologist himself, and wrote *La Langue Secrète des Dogons de Saga* (1948), while Griaule is well-known for his studies of the cosmology of the Dogon people in Mali.[63] Griaule and Dieterlen's work was aimed at the discovery of the *esoteric*, the cosmogonic secrets of the Dogon, another interesting theme in common with astrology.[64]

CLS developed his interest in surrealism from his close relations with André Breton and Max Ernst during their exile in New York, especially through the opportunities they had to buy Indian masks from the Pacific Northwest coast. One of CLS's last works was *The Way of the Masks*.[65] It

[60] Lévi-Strauss, 'Voix Off', p. 30.

[61] James Clifford, 'On Ethnographic Surrealism', *Comparative Studies in Society and History* 23, no. 4 (October 1981): pp. 539–564, p. 542.

[62] Wilcken, *The Poet*, p. 184; and Claude Lévi-Strauss, *The Savage Mind* (London: Weidenfeld and Nicolson, 1966), p. 17

[63] Michel Leiris, *La Langue Secrète des Dogons de Sanga* (Paris, Editions Jean-Michel Place, 1948). Marcel Griaule, *Conversations with Ogotemmeli: an introduction to Dogon Religious Ideas* (London: International African Institute, Oxford University Press, 1970).

[64] James Clifford, *The Predicament of Culture* (Cambridge, MA: Harvard University Press, 1988), p. 85.

[65] Lévi-Strauss, *The Way of the Masks*, trans. Sylvia Modelski (1982: Seattle: University of Washington Press, 1988).

can be argued that the interest of anthropologists in Amerindian and African art in turn influenced the surrealists, who were concerned to break out of what they regarded as the prison of conventional rationality.

And in fact there are very surreal passages in some traditional myths, as this extract shows. In *The Origin of Table Manners* Lévi-Strauss describes a Guianan myth (M524) about the origin of the anus.

> In the beginning neither men nor animals had anuses but excreted through their mouths. One day a disembodied anus sauntered among them, farting in their faces and then escaping. But they hunted him down, cutting him up in pieces and sharing him out among all animals – bigger or smaller, in accordance with the size of an orifice today.[66]

The crossover with anthropology has been examined by a number of writers, including James Clifford, James Hollier and J. Michael Dash. Hollier concluded that Clifford had exaggerated the connection, because except during the period of *Documents,* there was never a group of surrealist-anthropologists.[67] Dash draws attention to the important influence of surrealism on the Martinican poet and campaigner for black liberation, Aimé Césaire, which was reinforced by André Breton's arrival in Martinique in 1941, accompanied by Lévi-Strauss, and when Breton later returned and became involved in Haitian politics.[68] Pierre Mabille was another key figure in the Caribbean, and one who met André Barbault.[69] Mabille was a doctor, anthropologist and diplomat as well as being an astrologer.

In 1968 Lévi-Strauss gave an interview to Claude Courtot, for the surrealist magazine *L'Archibras*, during which he responded to the following questions:

> As the father of structuralism, are you happy with your offspring? Putting ethnography aside, what do you think of incest? As a Bororo tribesman, how

---

[66] Cited by Wilcken, *The Poet*, p. 292

[67] See Dennis Hollier, 'Surrealism and its Discontents', 2007, available at http://www.surrealismcentre.ac.uk/papersofsurrealism/journal7/acrobat%20files/articles/Hollierpdf.pdf [accessed 20 June 2016].

[68] J. Michael Dash, 'Le Je de l'autre: Surrealist Ethnographers and the Francophone Caribbean', *L'Esprit Createur* 47, no. 1 (2007): pp. 84–95.

[69] André Barbault, interviewed by Fabrice Pascaud in 2007, in *L'Astrologue* 161 (2007), Special Issue. This issue seems impossible to find in libraries, and the publishers did not respond to inquiries, but the whole interview is available in Italian in 3 parts on the website www.enzobarilla.eu

would you respond to a visit by M. Lévi-Strauss? Do you have any hopes for space travel? [70]

Lévi-Strauss entered into the spirit of the interview and replied to the second question in terms that no public figure would employ today. But none of this implies that Lévi-Strauss was willing to embrace *surrealisme sauvage* in the style of Georges Bataille and Michel Leiris: that was a step too far, as was belief in astrology. In 1954 Lévi-Strauss had been involved in a virulent controversy with Roger Caillois, one aspect of which was the latter's claim that Lévi-Strauss had been influenced by Dada and Surrealism, which Lévi-Strauss rejected in a crushing 30-page response in *Les Temps Modernes*.[71] And according to Patrick Wilcken, Lévi-Strauss became suspicious even of modern art movements including Cubism:

'while primitive art was a collective enterprise embedded in the societies in which it was produced and fused with their ritual and religious lives, Cubism was a contrived escape into an individualised aesthetic world'.[72]

Despite Barbault's friendship with Breton, in an interview in 2007 he said that he had not been much influenced by surrealism, although he then went on to tell an anecdote about Breton.[73] When Barbault had suggested to Breton, that astrological cycles involving Jupiter and Uranus as well as Saturn and Pluto were related to the development of Dada and hence surrealism, and to the first world war in 1914, Breton apparently rejected the idea, saying that Surrealism was not determined by the planets.[74] This makes an interesting contrast with Breton's own interest in the synodic cycle of Saturn and Uranus, which he noted in the birth charts of his most esteemed antecedents, Gérard de Nerval (22 May 1808) and Artur Rimbaud (20 October 1854), and contemporaries Louis Aragon (3 October 1897), Paul Éluard (14 December 1895) and himself (19 February 1896).

---

[70] Lévi-Strauss ,'Voix Off, Claude Courtot entretien avec Claude Lévi-Strauss', pp. 27–31.

[71] Loyer, *Lévi-Strauss*, pp. 404–408.

[72] Wilcken, *The Poet*, p. 236.

[73] Tessel M. Baudoin, *Surrealism and the Occult: Occultism and Western Esotericism in the Work of André Breton* (Amsterdam: Amsterdam University Press, 2014), pp. 109–111.

[74] Barbault, 'Entretien', p. 2, available at: www.enzobarilla.eu/Intervista_Barbault_Pascaud_II_Parte.pdf [accessed 1 July 2016].

Baudelaire, he noted, was born at the time of a Uranus-Neptune conjunction (9 April 1821). In two of these cases Saturn and Uranus were quite far apart, (Rimbaud 30 degrees, and Nerval 16 degrees), more than would be accepted as conjunctions by most astrologers. And if artists could be identified astrologically so easily we should include Joseph Goebbels (29 October 1897), who indeed obtained a PhD for a thesis on a romantic writer, but was known to detest modern culture and was a little lacking in surrealist humour.[75] Breton went to the length of changing his own birth date by one day, so as to have the Sun in Aquarius instead of Pisces, although there may also have been other personal reasons for this.[76]

A belief in the need for *defamiliarisation* is quite in tune with the surrealists' approach, and it was also an important feature of the Russian Formalist movement during the First World War, led by Viktor Shklovsky. Another member of this group was Roman Jakobson, who founded – in his first period of enforced exile – the Prague School of Linguistics. In New York Jakobson became one of the major acknowledged influences on Lévi-Strauss's linguistic structuralism applied to myth.[77]

Another feature of Breton's surrealist approach was the belief in attaining the unity of opposites through their art, which was also the purpose of myth in the view of Lévi-Strauss. This suggests a therapeutic aspect of surrealist art. And the surrealist belief in 'objective chance' by which the inner-outer dynamic was expressed through encounters that were pre-ordained rather than random, recalls the correspondences that are a central feature of astrology, as well as Jung's concept of synchronicity or an 'acausal connecting principle'.[78] Despite this, Jung, whose interests in astrology and alchemy were shared by the surrealists, was rejected by them as a reactionary and a fascist sympathiser. Freudian thought and especially Freud's view that sexuality was the basis of all action was still part of the central dogma of surrealism.[79]

---

[75] Baudoin, *Surrealism*, pp. 109–111. For Goebbels, see Roger Manvell and Heinrich Fraenkel, *Dr. Goebbels, his Life and Death* (New York: Skyhorse, 2010).
[76] Baudoin, *Surrealism*.
[77] Hawkes, *Structuralism*, pp. 60–73.
[78] Baudoin, *Surrealism*, p. 119. See also Nicholas Campion, 'Surrealist Cosmology, André Breton and Astrology', *Culture and Cosmos* 6, no. 2 (Autumn/Winter 2002): pp. 45–56, available at www.cultureandcosmos.org.
[79] Baudoin, *Surrealism*, p. 143.

Breton was the leader of the more rational wing of the movement, which valued academic study and even attempted to assess objective chance by distributing a questionnaire.[80]

Barbault knew the psychoanalyst René Allendy who was the source of his group's interest in connections between psychoanalysis and astrology 'at the end of WW2'.[81] This is interesting because over a decade earlier in 1933, Allendy had organised a series of lectures at the Sorbonne on '*pensee magique*', which may well have been catalytic in promoting the interest of the surrealists in the common ground with anthropology, based on dreams, magical thought and neuroses.[82] Breton's surrealist group's rapprochement with the occult did not really begin until about 1943, but their studies of works by Sir George Frazer and Lévy-Bruhl as well as Freud was certainly a pre-disposing force. Another important feature in tune with social anthropology was that the surrealists followed Lévy-Bruhl in rejecting the idea, espoused by Freud and Frazer, that primitive thought was inferior: to them however, unlike Lévi-Strauss, it was more than just different – it was superior to modern western thought.

This attitude, of course, derived from the surrealist desire for a revolution in consciousness in the west, whose reliance on rationality had, in their view, been responsible for the devastation of the 1914–18 war. Myth too became important to them, and in 1935 Breton gave two lectures in Prague in which he put forward the necessity for a new collective myth in western society, whose development would naturally be the task of the surrealists.[83] The journal *Minotaure* was the vehicle for many publications about myth, including two articles by Jacques Lacan.[84]

So, all in all, we can see family resemblances between surrealism, occultism, astrology, and nascent social anthropology, but also a lot of family squabbles, and in the end the relevance of surrealism to the scientific study of culture diminished, while remaining stronger in the arts and also in post-colonial politics in the Caribbean.

The motivations of the editor of *L'Astrologue* for contacting Lévi-Strauss are not easy to analyse, but he was part of a group of intellectuals in Paris during the 1960s which included psychologists – even Jacques

---

[80] Baudoin, *Surrealism*, p. 119.

[81] Barbault, *The Value*, p. 168 and footnote 81.

[82] Baudoin, *Surrealism*, p. 121.

[83] Baudoin, *Surrealism*, p. 123; and see André Breton, *Manifestoes of Surrealism*, trans. Richard Seaver and Helen R. Lane (Ann Arbor, MI: Ann Arbor Paperbacks, 1972).

[84] Baudoin, *Surrealism*, p. 126.

Lacan on one occasion – and artists, and he later interviewed André Breton for another issue of the same magazine.[85] At least one member of this circle, the astrologer and Tarot reader Fabrice Pascaud, was a member of the reconstituted group of surrealists during the period 1979–84, but he has not responded to requests for information. Another indication of the overlap of Barbault's interests with surrealism, despite denying its influence on his work, is an interview that he did in 1973 with Guy Michaud, a literature professor who has written about surrealism and published a book, *Le Visage Interieur: pour une anthropologie de l'ecrivain*, in which he used the planetary archetypes and personality psychology. In his interview with Barbault, Michaud states that he used astrology explicitly in his teaching.[86] It is interesting that Barbault attempts a similar classification to Michaud's, to whom he refers very briefly, but of painters instead of writers.[87]

The way astrologers work
At the end of the published interview Lévi-Strauss proposed a joint study of astrologers' works, with a view to understanding how they think. According to Dr Nicola the suggestion was not pursued, because 'there are as many astrologies as there are astrologers, and most of them are ego-centric and individualist'.[88] But in the last 20 years several theses have been completed which involve different approaches to exactly this question. I hope it will be useful to cite some examples.[89]

The thesis which looks most closely at what astrologers do in a consultation is Kirsten Munk's, where she quotes a large number of astrologers discussing their methods.

The well-known astrologer Liz Greene is quoted as saying she has no idea how astrology works, but views it as divination, and says 'Everything is connected, the universe is alive.' She also mention's Jung's concept of the quality of a moment in time.[90] Bernadette Brady, whose background is in biological research, talks of escaping scientific reductionism and says

---

[85] See *Intervista*, Parts 1 and 2, www.enzobarilla.eu [accessed 1 July 2016].
[86] Barbault 'Entretien avec Guy Michaud', *L'Astrologue* 21 (Spring 1973), pp.1–5.
[87] Barbault, *Value*, p. 232–249.
[88] Jean-Pierre Nicola, email to Graham Douglas, 21 May 2016.
[89] A number of such theses are available on Patrick Curry's website, available at www.the9thhouse.org.
[90] Kirstine Munk, *Signs of the Times, Cosmology and Ritual Practice in Modern Western Astrology* (PhD thesis, University of Southern Denmark, Odense, 2007), p. 154.

that 'astrology translates a dialogue with the cosmos'.[91] John Wadsworth emphasizes the ritual element of what he does, taking time to be well-prepared to enter a different state of consciousness during which he 'asks a spirit guide to watch over the consultation'.[92]

Again, Greene describes rare occasions when a deep rapport was established with a client and the transformative moment became very intense, saying that it happens very quickly with astrological symbols, in comparison to normal counselling work.[93] The astrologer Steven Forrest speaks of being overtaken by the archetypes which speak through him.[94] Munk then goes on to describe the effects of a consultation on clients with many transcripts from her conversations, as Vilhena did with astrology students in Brazil.[95]

In his discussions Vilhena noted a number of recurring themes when he inquired why students were interested in astrology, including a search for meaning, a need for spirituality free from the rigid constraints of either Catholicism or Freudian psychoanalysis, and as a means of developing and negotiating their identity in modern society. Some of his informants had experimented with Candomblé practices with varying degrees of commitment.

Clearly the practice of astrology has moved on from the typical 'find your personality and your destiny', and become more in tune with postmodern and post-structuralist thinking without ever incorporating these analyses into its textbooks. When we consider the way that astrologers have attempted to bring the chart features in line with the Jungian archetypes we have already seen that there is a good deal of confusion. As Gras points out, the number of archetypes is potentially limitless, so any correspondences with an astrological chart will not be convincing.[96]

So if Lévi-Strauss had been able to proceed with his suggested project of analysing astrological interpretations he would likely have found himself dealing with material closer to literature than to myth, a difficult situation for the Lévi-Strauss of 1969, as illustrated by the sardonic response he made to Roland Barthes' Structuralist criticism. Given the current saturation of psychological astrology with Jungian psychology one

---

[91] Munk, *Signs*, p. 162.

[92] Munk, *Signs*, p. 182–183.

[93] Munk, *Signs*, p. 194.

[94] Munk, *Signs*, p. 197.

[95] Vilhena, *World*, Chapter 3, pp. 87–122.

[96] Vernon W. Gras, 'Myth and the reconciliation of oppositions: Jung and Levi-Strauss', *Journal of the History of Ideas* 42, no. 3 (1981): pp. 471–488.

wonders how he would have responded to analysing the readings of astrologers.

And inevitably Lévi-Strauss would have found multiple astrologies, perhaps related to their social environments. In the postmodern context of Rio de Janeiro, Vilhena noted that while some embraced Jungian astrology as a liberation from rigid Catholic spirituality, others went the opposite way, towards *The Tradition* espoused by René Guénon.[97] This splitting pattern may be a general phenomenon; it calls to mind what Chude-Sokei observed in Jamaica, where Rastafarianism sought a return to roots culture, while Rap's techno-poetics embraced a Creolized future.[98]

In a commentary on Jung and Lévi-Strauss, Gras says:

> If it is true that we can never escape from the "hermeneutical circle" or from culture viewed as a "chain of signifiers" which will never break through to their referents, that unmediated Nature is a phantasm, that Culture is a man-made horizon wherein both I and the Other find and transform our identities, that Truth is a matter of convention and that there is no neutral "objective" ground which can adjudicate rival truth claims, then how does one initiate cultural change and decide between alternatives in order to legitimize them?[99]

He continues,

> One cannot legitimate a specific change by referring it to the call of Being or to the general exercise of human freedom, except in the rhetorical manner of every reformer and prophet in the past. If the distinction between myth and science, traditionally expressed as between the false and the true, no longer holds and scientific descriptions of Nature receive their legitimation only from cultural beliefs and practices, then the sciences of today become the myths of tomorrow. No longer do "privileged representations" exist anywhere. The only standard or guide left to us is pragmatic.[100]

And as Vilhena concludes: 'Contrary to the theoreticians of occultism, it can be seen how astrology truly becomes a vehicle that expresses and problematizes the tensions of [modern] values themselves, even while apparently denying them'.[101]

---

[97] Vilhena, *World*, pp. 164–165.
[98] Louis Chude-Sokei, *The Sound of Culture: Diaspora and Black Techno-Poetics*, (Middletown, CT: Wesleyan University Press, 2016), pp. 143, 161, 181.
[99] Gras, *Myth*, p. 487.
[100] Gras, *Myth*, p. 487.
[101] Vilhena, *World*, p. 187.

Reviews

Dennis Danielson, *Paradise Lost and the Cosmological Revolution*
(Cambridge: Cambridge University Press, 2014). ISBN: 978-1107033603.
246 pp.

Dennis Danielson's *Paradise Lost and the Cosmological Revolution*
discusses the changes in cosmological modelling wrought by the new
schools of astronomers in the sixteenth and seventeenth centuries,
unpacking the implications of their observations and speculations for the
dialogues and cosmic portraiture of John Milton's *Paradise Lost*.
Danielson is Professor of English at the University of British Columbia,
and has long been a formidable Miltonist, with important books like
Milton's Good God (1980) and *The Cambridge Companion to Milton*
(1989), and a writer on the history of cosmology with essays like 'The
Great Copernican Cliché' (2001), an anthology of cosmological writings
(*The Book of the Cosmos*, 2000), and a biography on Georg Joachim
Rheticus (*The First Copernican*, 2006). *Paradise Lost and the
Cosmological Revolution* is a culmination bringing together Danielson's
two major scholarly interests from his earlier and more recent work: the
English epic poet and the history of cosmology. Notably, Danielson's book
occurs in the wake of substantial scholarly works on the new astronomy
and Milton, particularly Malabika Sarkar's *Cosmos and Character in
Paradise Lost* (2012), John Leonard's *Faithful Labourers* (2013), and
Karen Edwards' article in the *Cambridge Companion to Paradise Lost*
(2014), all reacting to centuries of Milton critics who thought of Milton as
an undecided semi-Ptolemaist.

Danielson begins by discussing the demands of pre-Copernican
cosmology ('uniform, circular motion, upon crystalline spheres, turning
about a universal center point, which, for reasons of heaviness, is occupied
by an immobile Earth') and the basic diagrammatic elements, as well as
how early astronomers made astronomical predictions and judgments with
it. The remainder of the book is a systematic demonstration of how
Milton's depiction of the cosmos is not Ptolemaic. Danielson launches
these chapters with a discussion of Milton's perspectives on creation and
matter, indicating the extraneousness to 'our' universe of Milton's chaos
and hell, and the possibility of a multiverse in *Paradise Lost*. Afterwards,
he treats 'cosmological bricoleurs' – sixteenth century astronomers
(Copernicus, Brahe, Digges, and others – considering early theories of
heliocentrism, writings on the supernova of 1572, magnetism and geostatic

theories which nevertheless allowed diurnal rotation. The next two chapters make connections between Milton and Galileo, exploring these writers' shared delight in mutability and generation, provide a presentation of reflected light from the earth and a universe of dispersed stars, each with potentially new centres, as well as a concern with the relative magnitude and importance of the sun and the earth among the units of the universe. The next chapter considers sunspots and the theological and cosmological parallel between Milton's and Kepler's ideas and attitudes concerning the sun, especially on magnetism and solar glorification. The final two chapters ask '(1) What kind of place is Earth and (2) what kind of place is the Universe?' (p. 155). The first indicates the new cosmology's (and Milton's) appreciation of stellar earth, newly adopted into the dance of the stars, and the second addresses astronomers and Milton's speculation on other life in outer space and the nature and possibility of movement outside of earth.

If I have laid out a straightforward organization of Danielson's book, there is considerable obliquity and piecemeal in his approach. When one reads a book on Milton and astronomy, one expects prompt and straightforward explication of Raphael's discourse on astronomy from Book VIII of *Paradise Lost*. However, Danielson waits until near the end of his book (Chapter 7) for a first extensive treatment of this passage from the poem. In this way, Danielson thwarts easy and quick answers, and builds his argument with thick bonds of contextual material to an eventual argumentative climax. Chapter 2 is something of a readerly surprise, introducing a topic which is initially difficult to relate to the first chapter and the enterprise laid out in the preface and book jacket. Moreover, chapter two almost exclusively addresses Milton, while chapter three does not even mention Milton, so Danielson's method and synthesis function differently throughout the monograph. His treatment of Galileo gets two specific chapters, but arguably the most sumptuous connection to Milton is made in Chapter 7 (on planet earth). One hoping to understand the extent to which *Paradise Lost* was informed by Galilean perspectives might find the titles of these chapters a bit cryptic: 'Milton and Galileo revisited (1): "Incredible delight"' and 'Milton and Galileo revisited (2): "What if?"'). Moreover, they may press on only to be dizzied by quite extensive discussions about Francis Bacon, and then about how Milton's hesitation to endorse one cosmological system explicitly over another has perhaps less to do with the intellectual equality of the Ptolemaic model than with the strength and currency of Tycho Brahe's ideas. Although moderate in length, not all of the book is easeful reading, and the chapter headings and epigraphs do not fully prepare one for the range of topics or authors covered in some chapters. Danielson is interested in bringing strong nuance to the critical discussion of Milton, which sometimes leads him to make scrupulous distinctions and introduce obscure texts, and at other times, produce straightforward propositions about well-known and utterly canonical texts. His introduction to the Ptolemaic cosmos in Chapter 1, as well as and his constant 'asides' to the twenty-first century reader (usually alerted by Danielson's frequent use of 'still today' or 'even today') sometimes feel at odds with his subsequent name-dropping and subtle differentiations between Renaissance cosmographies.

My chief complaint with his book is that it does not consider astrology ('astrology' is not even indexed), but seems only to use the monolithic term more acceptable to modern sensibilities, 'astronomy'. Astrological predictions were the motivation for most 'astronomical' work, and in reality there was limited distinction between the two. The nature, function and importance of the zodiac are not mentioned at all in *Paradise Lost and the Cosmological Revolution*, in spite of making an appearance in *Paradise Lost*. The silence on astrology is particularly surprising given that the first chapter is a namesake of C. S. Lewis's *The Discarded Image*, a text which considers astrology at some length. Danielson has numerous other opportunities to bring up astrology: in his discussion of planetary influence and the reality of earthshine, in his bringing up of Platonism (but eliding the hermetic traditions) and even famous astrologers like Marsilio Ficino. Perhaps Danielson's book moves along more smoothly without diverting to consider astrology, but the result feels a tad omissive.

In the main, Danielson's arguments are shatteringly strong and the connections he makes perspicuous and unforgettable for any re-reading of *Paradise Lost*; so much so, that to think that the counter-view held a moderate consensus in the past is baffling. Furthermore, a noble endeavour throughout Danielson's book is an unrelenting critique of the 'smug hindsight' associated with a 'unidirectional cosmological revolution' that does not see 'the necessity of piecemeal model-making' (p. 74). Particularly strong are Chapters 7 and 8 (on planet earth, ET, and space travel), which deal with topics almost untouched by Milton critics. Danielson's sources are numerous but uncumbersome, and his passionate debunking of popular myths results in a delightfully polemic tone, which transitions into an exuberant show-and-tell of the possible in Chapter 8. Considering the book at large, one gets the sense that Danielson has researched multifarious scholarly debates on the many matters which the book touches, though he is not overbearing with footnotes, and reserves quotation almost exclusively to primary texts, which agrees spiritually with the remarkably brief 'Bibliographical note' that concludes the book: 'If one wishes to explore the history of astronomy and cosmology further, one cannot do better than to sample the writings of the great astronomers themselves: Copernicus, Tycho Brahe, Kepler, Galileo' (p. 215). In summary, *Paradise Lost and the Cosmological Revolution* should be a permanent must-read for any serious Milton critic, and a prime text for Renaissance scholars interested in seeing how the cosmological changes wrought in the sixteenth and seventeenth centuries informed literary productions.

Richard Bergen, University of British Columbia

Luís Rodolfo Vilhena, *The World of Astrology: An Ethnography of Astrology in Contemporary Brazil*, trans. Graham Douglas (Ceredigion: Sophia Centre Press, 2014). ISBN: 978-1-907767-04-3. Illustrated, 244 pp.

The author of this book, Luís Rodolfo Vilhena, was a promising Brazilian anthropologist who died tragically young in 1997 at the age of thirty-three. *The World of Astrology*, based upon his research for a Masters degree at the University of Rio de Janeiro, was originally published in Portuguese in 1990. Its chance discovery (as we say) by Graham Douglas in a Lisbon bookshop inspired him to produce this excellent translation, and both he and the Sophia Centre Press are to be congratulated for the resulting new addition to the Anglophone world of scholarship and research into modern astrology.

Douglas also contributes a helpful preface in which he situates Vilhena's work in a double context: influences on that work, especially Claude Lévi-Strauss, and subsequent research in English conducted independently, especially by Alie Bird, Kirsten Munk, Bridget Costello, Bernadette Brady and Nicholas Campion. Their work comprises a mixture of ethnography, anthropology more broadly, and sociology.[1]

Vilhena's subjects are members of the urban middle classes in Rio de Janeiro with varying degrees of involvement in astrology, from professional practitioners to those who only consult astrologers. They are also involved with astrology in ways and for reasons which differ. The

[1] Much of it, although by no means all, is available at http://www.the9thhouse.org/index.htm [accessed 29 September 2016].

period is the 1980s, but surprisingly little seems to have changed. Some informants value astrology as a spiritual path (although not formally religious), some as a psychotherapeutic practice allied with Jung's analytical psychology, and some as an esoteric knowledge resisting the scientific materialism of modernity. The only thing missing here is the subsequent rise of astrology as divination which, because it doesn't fall neatly into any of those categories, has complicated them in an interesting and potentially fruitful way.

The strength of Vilhena's approach follows from his adherence to Lévi-Strauss's structuralism, which reveals the scope, sensitivity and flexibility of astrology as a classificatory system, based on synchronic binary oppositions, with which to make sense of experience, social relationships and the world. The ultimate development of this kind of astrology is perhaps in the orientations it enables towards the modern world as such, in tandem with the way its academic study can reveal those orientations.

Vilhena shows convincingly, for example, that rather than rejecting science outright, some of his astrologers are trying to spiritualise it. Others are working to the same end using psychology as a project that is, for them, both scientific and spiritual. That was precisely Jung's hope, of course. (The result can equally be seen as a disingenuous attempt to disguise its real nature, a muddled but pragmatic compromise, or a promising new synthesis.) Still others reject modern materialism altogether, taking refuge in astrology as an ancient esoteric and occult 'science' of the kind defended by the rebarbative René Guénon. But as Vilhena points out, both that rejection and the presumptive remedy are themselves thoroughly modern responses.

Vilhena makes the related point (as have others) that astrology's emphasis on exact astronomical positions, mathematical calculations and a complex set of theoretical rules for interpretation potentially position it as a scientific and/or objective enterprise, while the irreducibility of qualitative planetary principles, never far removed from divinities, equally mark it as 'magical'. Again, it offers, or seems to offer, a solution to the question of how to be in the modern world but not of it.

It seems worth adding that magic in fact offers a deeply compromised way to oppose the modern world. Although 'spiritual', a great deal of it is already implicated in the mode of instrumental power-knowledge that is so central to modernity: aiming for mastery, manipulating 'energies', using the will to bring about desired changes (whether 'subjective' or 'objective'). That which is radically non-modern, and which therefore marks its limits, is something else: enchantment. (Although, confusingly, the same word – magic – is often used to describe it). But wonder cannot be used, let alone organised, and with that realisation, people who mainly want power lose interest.[2]

Vilhena works hard to relate the various positions taken to the social classes and relationships of their takers, and with some success. It's odd, though, that he doesn't seem to have been aware of T. W. Adorno's early and influential writings on astrology based on the *L.A. Times'* sun-sign column and its readers in the 1950s.[3] Amid Adorno's dollops of Marxism and psychoanalysis and his wildly speculative conclusions are some valuable insights, especially the idea that astrology, in any depth beyond sun-sign columns, appeals mostly to the 'semi-erudite'. By this, Adorno meant those sufficiently well educated to follow its sometimes intellectually demanding complexities, but not so well-educated that they have thereby succumbed to the intellectual elite's metaphysical worldview. (Since the late seventeenth century, that has been one which excludes even the possibility that astrology is true or real.) Of course, this concept is also too crude, but it is at least interesting and potentially fruitful.

Although it's not a serious omission, the commentary here might also have mentioned Bauer and Durant's 1997 empirical study 'Belief in Astrology', which follows up Adorno's work. It broadly supports the conclusions in this book.[4]

Not surprisingly, the weakness of Vilhena's work also follows from the source of his insights, namely its structuralism. That commitment means, as he says, that 'I approached astrology as a whole principally in terms of its beliefs' (p. 103). Belief and knowledge are functions of epistemology. As such, they encourage a neglect of how astrology works as ontology: a way of life, not only a way of knowing, in which working with symbolism, arguably the heart of astrology, is central. We learn much about various worldviews and their social dimension, but it is possible to miss a close

---

[2] I explore this and related issues in 'Enchantment and Modernity', *PAN: Philosophy, Activism, Nature* 9 (2012): pp. 76–89, available at http://arrow.monash.edu.au/vital/access/manager/Repository/monash:85446 [accessed 29 September 2016].
[3] T. W. Adorno, 'The stars down to earth: The Los Angeles Times astrology column, a study in secondary superstition', *Jahrbuch für Amerikastudien* 2 (1957): pp. 19–88, reprinted in T. W. Adorno, *The stars down to earth and other essays on the irrational in culture*, ed. Stephen Crook (London and New York: Routledge, 1994).
[4] Martin Bauer and John Durant, 'Belief in Astrology: A Social-Psychological Analysis', *Culture and Cosmos* 1 (1997): pp. 55–72.

study of how astrological symbolism itself works when it is an essential part of lived experience. For that – not as a replacement for Vilhena's sociology and social anthropology, but as a necessary complement – a more phenomenological and/or hermeneutic approach is necessary.

Such a call by no means rules out anthropology, which is a very roomy (and contested) discipline. It does, however, move in the direction of the humanities and away from the social sciences. A start, and good example, is provided by an MPhil thesis briefly mentioned in Douglas's preface: Lindsay Radermacher's 'The Role of Dialogue in Astrological Divination' (2011).[5]

It also follows, I think, that to understand what it's like to be a practising astrologer (including, but not only, what it feels like), one needs to at least have had the experience of being one.[6] Vilhena studied astrology but mainly, it seems, as a 'system' which one 'applies' to generate meaning. It is that, and an admirable and fascinating one, as this fine study shows; but it is far from only that.

Patrick Curry, University of Wales Trinity Saint David

---

[5] Lindsay Radermacher, 'The Role of Dialogue in Astrological Divination' (MPhil, University of Kent, 2011), available at http://www.the9thhouse.org/docs/Lindsay%20Radermacher%20MPhil%20Thesis%202011.pdf [accessed 29 September 2016].

[6] I address this issue in relation to the history of astrology in 'The Historiography of Astrology: A Diagnosis and a Prescription', in *Horoscopes and Public Spheres: Essays on the History of Astrology*, ed. K. von Stuckrad, G. Oestmann, and H. D. Rutkin (Berlin and New York: Walter de Gruyter, 2005), pp. 261–74.

NOTES ON CONTRIBUTORS

Helena Avelar holds an MA in Medieval History from the Faculty of Social Sciences and Humanities of the Nova University of Lisbon. She is a researcher at CIUHCT – Centro Interuniversitário de História das Ciências e da Tecnologia, and at IEM - Institute of Medieval Studies. Her research focuses on medieval science and culture, with a special interest in astrological practices and techniques. She is currently engaged in PhD studies at the Warburg Institute in the University of London, under the supervision of Professor Charles Burnett.

Roger Beck is Professor Emeritus at the University of Toronto, where he began his career at Erindale College and Department of Classics (Lecturer 1964-65, Assistant Professor 1968-74, Associate Professor 1974-84, Professor 1984-98, Professor Emeritus since 1998). He received his BA from Oxford (1961) and his PhD in Classical Philology at University of Illinois (1971). He was Secretary of the Classical Association of Canada from 1977-79 and Review Editor (1978-82) and Associate Editor (1982-86) for the journal *Phoenix*. His current research interests are Mithraism and religion in the Roman Empire; ancient astrology and astronomy; and Petronius and the ancient novel. His works include *The Religion of the Mithras Cult in the Roman Empire* (2006) and *A Brief History of Ancient [this word added 2/16] Astrology* (2006). His collected articles, many of them having to do with ancient astronomy and astrology, were published by Ashgate in 2004 as *Beck on Mithraism: Collected Works with New Essays*.

Richard Bergen is an English PhD candidate at the University of British Columbia, Canada. He works with late medieval and early modern non-dramatic poetry and prose, and has strong research interests in allegory, genre theories, cosmology, as well as space and place. He regularly presents on these topics, and has published an article on genre theory and the nature of John Bunyan's usage of allegory. He holds a Social Sciences and Humanities Research Council Canada Graduate Scholarship, and achieved the Mary Henley scholarship for being the top Renaissance student at UBC (2015).

Charles Burnett, MA, PhD, LGSM, FBA, is Professor of the History of Arabic/Islamic Influences in Europe at the Warburg Institute, University of London, and Co-Director of the Centre for the History of Arabic Studies in Europe at the Warburg Institute. His research centres on the transmission of texts, techniques and artefacts from the Arab world to the West, especially in the Middle Ages. He has documented this transmission by editing and translating several texts that were first translated from Arabic into Latin, and also by describing the historical and cultural context of these translations. Throughout his research and his publications he has aimed to document the extent to which Arabic authorities and texts translated from Arabic have shaped European learning, in the universities, in medical schools and in esoteric circles. Among his books in this subject area are *The Introduction of Arabic Learning into England* (1997), *Arabic into Latin in the Middle Ages: The Translators and their Intellectual and Social Context* (2009) and *Numerals and Arithmetic in the Middle Ages* (2010).

Patrick Curry holds a PhD in the History and Philosophy of Science, has taught at the universities of Bath Spa and Kent, and is a Tutor in the Sophia Centre for the Study of Cosmology in Culture at the University of Trinity Saint David. He is the co-author (with Roy Willis) of *Astrology, Science and Culture* (Berg, 2004) and editor of *Divination* (Ashgate/ Routledge, 2010).

Graham Douglas studied Chemistry at Imperial College London and worked in chemical and biophysical research for a number of years before becoming a schoolteacher. He has been interested in astrology since 1977, both for the empirical challenge it offers for investigating the evidence for possible geomagnetic influences on the processes of conception and birth, and also as a topic in cultural and semiotic research. His empirical studies, based on the data collected by Michel and Francoise Gauquelin, have been published in the journal *Correlation*, and on the website http://Cura.free.fr, the most recent of which is 'The Gauquelin Effect is born at Conception'. The academic journal *Semiotica* has published three of his articles dealing with astrology and semiotics, especially the work of A. J. Greimas. Since 2010 he has worked as a translator of Spanish and proofreader for the online bilingual and multicultural newspaper www.theprisma.co.uk, where he has also contributed articles and interviews, especially of film directors. In 2014 his translation of the book *O Mundo da Astrologia* by the Brazilian researcher Luís Rodolfo Vilhena, was published as *The World of Astrology* by the Sophia Press. Vilhena's ethnography of astrology is also one of the sources for his discussion of the interview with Lévi-Strauss.

176

Scott E. Hendrix graduated from the University of Tennessee in 2007 with a Ph.D. in Medieval history, specializing in intellectual history. He is the author or editor of six books, and his first study-- *How Albert the Great's Speculum Astronomiae Was Interpreted and Used by Four Centuries of Readers*—won the Professor D. Simon Evans prize for medieval studies. He has also published numerous articles and book chapters on topics such as the history of astrology, mysticism, the witch craze, and contextual rationality. He is currently an Associate Professor of History at Carroll University in Waukesha, Wisconsin (USA).

R. Hakan Kirkoğlu has studied astrology since 1983 and is currently working on his MA thesis titled 'Ilm-i Nudjum and its role in the Ottoman court during the eighteenth century' at Boğaziçi University in Istanbul. He holds an MA degree in Economics (1991) from Boğaziçi University and a BSc in Management Engineering (1988) from Istanbul Technical University. He is a consultant astrologer, writer and tutor.

Lindsay Starkey is an Assistant Professor of History at Kent State University at Stark. She received her PhD from the University of Wisconsin-Madison in 2012. Her research focuses on the ways in which sixteenth-century Christian assumptions about God's creation and providential guiding of the world influenced European understandings of the universe. She is currently working on a manuscript about sixteenth-century European conceptions of water.

BACK ISSUES OF CULTURE AND COSMOS
http://www.cultureandcosmos.org/backIssues.html

178

The Cross in Cygnus; **Angela Voss:** *The Astrology of Marsilio Ficino: Divination or Science?;* **Patrick Curry:** *Astrology on Trial, and its Historians: Reflections on the Historiography of 'Superstition'.*

Contents Vol. 5 no 1 (spring/summer 2001)
Demetra George: *Manuel I Komnenos and Michael Glykas: A Twelfth-Century Defence and Refutation of Astrology,* Part I; **Richard L. Poss:** *Stars and Spirituality in the Cosmology of Dante's* Commedia.

Contents Vol. 5 no 2 (autumn/winter 2001)
Arkadiusz Sołtysiak: *The Bull of Heaven in Mesopotamian Sources;* **Demetra George:** *Manuel I Komnenos and Michael Glykas: A Twelfth-Century Defence and Refutation of Astrology,* Part 2; **Garry Phillipson** and **Peter Case:** *The Hidden Lineage of Modern Management Science: Astrology, Alchemy and the Myers-Briggs Type Indicator.*

Contents Volume 6 Number 1 (spring/summer 2002)
Ari Belenkyi: *A Unique Feature of the Jewish Calendar - Deḥiyot;* **Demetra George:** *Manuel I Komnenos and Michael Glykas: A Twelfth-Century Defence and Refutation of Astrology,* Part 3; **Germana Ernst:** *The Sky in a Room: Campanella's Apologeticus in defence of the pamphlet* De siderali fato vitando; **Tommaso Campanella:** *Apologia for the opuscule on* De siderali fato vitando.

Contents Volume 6 Number 2 (autumn/winter 2002)
Jesse Krai: *Rheticus' Poem* 'Concerning the Beer of Breslau and the Twelve Signs of the Zodiac'; **Anna Marie Roos:** *Israel Hiebner's Astrological Amulets and the English Sigil War;* **Nicholas Campion:** *Surrealist Cosmology: André Breton and Astrology.*

Contents Volume 7 Number 1 (spring/summer 2003) GALILEO'S ASTROLOGY
Nick Kollerstrom: *Foreword: Galileo as Believer;* **Nicholas Campion:** *Introduction: Galileo's Life and Work;* **Antonio Favaro:** *Galileo, Astrologer;* **Germana Ernst:** *Astrology and Prophecy in Campanella and Galileo;* **Nick Kollerstrom;** *Galileo as an Astrologer: Antonino Poppi: On Trial for Astral Fatalism: Galileo Faces the Inquisition;* **Guiseppe Righini:***Galileo's Horoscope for Cosimo II de Medici;* **Mario Biagioli:** *An Astrologico-Dynastic Encounter; Galileo's Correspondence; Galileo's Letter to Dini, May 1611; On the Character of Sagredo: Galileo's judgements upon his nativity; Galileo's Horoscopes for his Daughters; Rome, 1630;* **Bernadette Brady:** *Four Galilean Horoscopes: An Analysis of Galileo's Astrological Techniques; A Sonnet by Galileo.*

Contents Volume 7 Number 2 (autumn/winter 2003)
Günther Oestmann: *Tycho Brahe's Geniture;* **Bernard Eccles:** *Astrological physiognomy from Ptolemy to the present day;* **James Brockbank:** *Planetary signification from the second century until the present day;* **Julia Cleave:** *Ficino's Approach to Astrology as Reflected in Book VII of his Letters.*

Contents Volume 8 No 1/2 (spring/summer autumn/winter 2004)
Valerie Shrimplin *Organising INSAP;* **Rolf Sinclair** *Foreword: INSAP IV in Oxford: A Summary;* **Nicholas Campion** *Introduction: The Inspiration of Astronomical Phenomena;* **Hubert A. Allen, Jr.** *Hawkins' Way: Remembering Astronomer Gerald S. Hawkins;* **Hubert A. Allen, Jr. and Terry Edward Ballone** *Star Imagery in Petroglyph National*

Monument; **Mark Butterworth** *Astronomy and the Magic Lantern*; **Ann Laurence Caudano** *Sun, Moon, and Stars on Kievan Rus Jewellery (10th – 13th Centuries)*; **Nicholas Campion** *The Sun is God;* **Anne Chapman-Rietschi** *Cosmic Gardens*; **Deborah Garwood** *Paris Solstice*; **N. J. Girardot** *Celestial Worlds In the Work of Self-Taught Visionary Artists With Special Reference to Howard Finster's Vision of 1982*; **John G. Hatch** *Desire, Heavenly Bodies, and a Surrealist's Fascination with the Celestial Theatre*; **Holly Henry** *Bertrand Russell in Blue Spectacles: His Fascination with Astronomy*; Ronald Hicks *Astronomy and the Sacred Landscape in Irish Myth*; **Chris Impey** *Why Are We So Lonely?*; **Bernd Klähn** *The Aberration of Starlight and/in Postmodernist Fiction*; **Nick Kollerstrom** *How Galileo dedicated the moons of Jupiter to Cosimo II de Medici*; **Arnold Lebeuf** *Dating the five Suns of Aztec cosmology*; **Andrea D. Lobel** *Trailing the Paper Moon: Astronomical Interpretations of Exodus 12:1-2*; **Stephen C. McCluskey** *Wordsworth's 'Rydal Chapel' and the Astronomical Orientation of Churches*; **David Madacsi** *Sky: Atmospheres and Aesthetic Distance in Planetary and Lunar Environments*; **Daniel R. Matlaga** *A Journey of Celestial Lights: The Sky as Allegory in Melville's Moby Dick*; **Paul Murdin** *Representing the Moon*; **R. P. Olowin** *Robinson Jeffers: Poetic Responses to a Cosmological Revolution*; **David W. Pankenier** *A Brief History of Beiji (Northern Culmen)*; **Richard Poss** *Poetic Responses to the Size of the Universe: Astronomical Imagery and Cosmological Constraints*; **Barbara Rappenglück** *The material of the solid sky and its traces in cultures*; **Brad Ricca** *The Night of Falling Stars: Reading the 1833 Leonid Meteor Storm*; **Patricia Ricci** *Lux ex Tenebris: Etienne-Louis Boullée's Cenotaph for Sir Isaac Newton*; **Sarah Richards** *Die Planetentheorie: its uses and meanings for the Saxon mining communities and the culture of the Dresden Court 1553-1719*; **William Saslaw and Paul Murdi**n *The Double Apollos of Istrus*; **Petra G. Schmidl** *Dusk and Dawn in Medieval Islam; On the Importance of Twilight Phenomena with Some Examples of Their Representations in Texts and on Instruments*; **Valerie Shrimplin** *Borromini and the New Astronomy: the elliptical dome*; **Joshua Stein** *Cicero's Use of Astronomy as Proof of the Existence of the Gods*; **Antje Steinhoefel** *Art and Astronomy in the Service of Religion:Observations on the Work of John Russell (1745-1806)*; **Burkard Steinrücken** *An interpretation of the `Sky Disc of Nebra' as an icon for a bronze age planetarium mechanism with parallels to the moving world-soul in Plato's* Timaeus; **Gary Wells** *Daumier and The Popular Image of Astronomy.*

Contents Vol. 9 no 1 (Spring/Summer 2005)
Gennadij Kurtik and Alexander Militarev *Once more on the origin of Semitic and Greek star names:an astronomic-etymological approach updated*; **Prudence Jones** *A Goddess Arrives: Nineteenth Century Sources of the New Age Triple Moon Goddess*; **Louise Curth** *Astrological Medicine and the Popular Press in Early Modern England.*

Contents Vol. 9 no 2 (Autumn/Winter 2005)
Marinus Anthony van der Sluijs *A Possible Babylonian Precursor to the Theory of ecpyrōsis*; **Liz Greene** *Did Orphic Beliefs Influence the Development of Hellenistic Astrology?*; **Ariel Cohen** *Astronomical Luni-Solar Cycles and the Chronology of the Masoretic Bible*; **Tayra Lanuza-Navarro** *An Astrological Disc from the Sixteenth Century*; **J.C. Holbrook** *Celestial Navigators and Navigation Stories.*

Contents Vol. 10 no 1 and 2 (Spring/Summer, Autumn/Winter 2006)
Lucia Dolce *Introduction: The worship of celestial bodies in Japan: politics, rituals and icons*; **Lucia Dolce** *The State of the Field: A basic bibliography on astrological cultic*

180

practices in Japan; **Hayashi Makoto** *The Tokugawa Shoguns and Yin-yang knowledge (onmyōdō)*; **John Breen** *Inside Tokugawa religion: stars, planets and the calendar-as-*method; **Mark Teeuwen** *The imperial shrines of Ise:An ancient star cult?*; **Lilla Russell-Smith** *Stars and Planets in Chinese and Central Asian Buddhist Art from the Ninth to the Fifteenth Centuries*; **Matsumoto Ikuyo** *Two Mediaeval Manuscripts on the Worship of the Stars from the Fujii Eikan Collection*; **Tsuda Tetsuei** *The Images of Stars and Their Significance in Japanese Esoteric Buddhist Art*; **Meri Arichi** *Seven Stars of Heaven and Seven Shrines on Earth: The Big Dipper and the Hie Shrine in the Medieval* Period; **Gaynor Sekimori** *Star Rituals and Nikko Shugendō*; **Meri Arichi** *The front cover image: Myōken Bosatsu.*

Contents Vol. 11 no 1 and 2 (Spring/Summer, Autumn/Winter 2007)
Micah Ross *A Survey of Demotic Astrological* Texts; **Francis Schmidt** *Horoscope, Predestination and Merit in Ancient Judaism*; **Stephan Heilen** *Ancient Scholars on the Horoscope of Rome*; **Joanna Komorowska** *Philosophy among Astrologers* ; **Wolfgang Hübner** *The Tropical Points of the Zodiacal Year and the* Paranatellonta *in Manilius'* Astronomica; Aurelio Pérez Jiménez *Hephaestio and the Consecration of Statues*; **Robert Hand** *Signs as Houses (Places) in Ancient Astrology*; **Dorian Gieseler Greenbaum** *Calculating the Lots of Fortune and Daemon in Hellenistic Astrology*; **Susanne Denningmann** *The Ambiguous Terms* ἑῴα *and* ἑσπερία, ἀνατολή, *and* ἑῴα *and* ἑσπερία δύσις **Joseph Crane** *Ptolemy's Digression: Astrology's Aspects andMusical Intervals*; **Giuseppe Bezza** *The Development of an Astrological Term – from Greek* hairesis *to Arabic* hayyiz; **Deborah Houlding** *The Transmission of Ptolemy's Terms: An Historical Overview, Comparison and Interpretation.*

Contents Vol. 12 no 1 (Spring/Summer 2008)
Liz Greene *Is Astrology a Divinatory System?*; **James Maffie** *Watching the Heavens with a 'Rooted Heart': The Mystical Basis of Aztec Astronomy*; **J.C. Holbrook** *Astronomy and World Heritage.*

Contents Vol. 12 no 2 (Autumn/Winter 2008)
Mark Williams *Astrological Poetry in late medieval Wales: the case of Dafydd Nanmor's 'To God and the planet Saturn'*; **Scott Hendrix** *Choosing to be Human: Albert the Great on Self Awareness and Celestial Influence*; **Graham Douglas** *Luis Vilhena and the World of Astrology.*

Contents Vol. 13 no 1 (Spring/Summer 2009)
Josefina Rodríguez-Arribas *Astronomical and Astrological Terms in Ibn Ezra's Biblical Commentaries: A New Approach*; **Andrew Vladimirou** *Michael Psellos and Byzantine Astrology in the Eleventh Century*; **Marinus Anthony van der Sluijs** *The Dragon of the Eclipses—A Note*; **Patrick Curry** *Response to Liz Greene's 'Is Astrology a Divinatory System?'*

Contents Vol. 13 no 2 (Autumn/Winter 2009)
Liz Greene *Mystical Experiences Among Astrologers*; **Peter Pesic** *How the Sun Stood Still: Old English Interpretations of Joshua and the Leap Year*; **Doina Ionescu** *Virginia Woolf and Astronomy*; **Carlos Ziller Camenietzki and Luis Miguel Carolino** *Astrologers at War: Manuel Galhano Lourosa and the Political Restoration of Portugal, 1640–1668*; **Nick Campion** *Astrology's Role in New Age Culture: A Research Note*

Culture and Cosmos

182

Siberia; **Christian Etheridge**, *A systematic re-evaluation of the sources of Old Norse astronomy*; **Aidan Foster**, *Hierophanies in the Vinland Sagas: Images of a New World*; **Inga Elmqvist Söderlund**, *Inspiration from antique heroic deeds: Hercules as an astronomer*; **Patricia Aakhus**, *Astral Magic and Adelard of Bath's Liber Prestigiorum; or Why Werewolves Change at the Full Moon*; **David Pankenier**, Astrology for an Empire: The 'Treatise on the Celestial Offices' (ca. 100 BCE); **Steven Renshaw**, *The Inspiration of Subaru as a Symbol of Values and Traditions in Japan*;b **Daniel Armstrong**, *Citing The Saucers: Astronomy, UFOs and a persistence of vision*; **Alberto Cappi**, *The concept of gravity before Newton*; **Paul Murdin**, *Artilleryman to head of state—how astronomy inspired Francois Arago*; **Paolo Molaro and Alberto Cappi**, *Edgar Allan Poe's cosmology in Eureka*; **Voula Saridakis**, *For 'the present and future happiness of my dear Pupils'": The Astronomical and Educational Legacy of Margaret Bryan*; **Michael Rowan–Robinson**, *The invisible universe*; THE ARTS: **Arnold Wolfendale**, *The Inter-Relation of the Visual Arts and Science in General and Astronomy in Particular*; **Lynda Harris**, *Changing Images of the Milky Way during the Greco-Roman and Medieval Periods*; **Lucia Ayala**, *The Universe in images: Iconography of the Plurality of Worlds*; **Tayra M. Carmen Lanuza-Navarro**, *Astrological culture before its public: the representation of astrology in Golden Age Spanish Theatre*; **Emily Urban**, *Depicting the Heavens: The Use of Astrology in the Frescoes of Rome*; **Michael Mendillo**, *The Artistic Portrayal of the Medicean Moons in Early Astronomical Charts, Books and Paintings*; **Rolf Sinclair**, *Howard Russell Butler: Painter Extraordinary of Solar Eclipses*; **Beatriz Garcia, Estela Reynoso, Silvina Pérez Alvarez and Rubén Gabellone**, *Inspiration of Astronomy in the movies: a history of a close encounter*; **Gary Wells**, *The Moon in the Landscape: Interpreting a Theme of 19th Century Art*; **Clive Davenhall**, *The Space Art of Scriven Bolton*; **Matthew Whitehouse**, *Astronomical Organ Music*; **Aaron Plasek**, *Between Scientists, Writers and Artists: Theorising and Critiquing Knowledge-Production at the Interstices between Disciplines*; ARTISTS: **Merja Markkula**, *The Way I See the Stars: fibre art inspired by astrobiology*; **Govinda Sah**, *Beyond the Notion*; **Gisela Weimann**, *Above all the stars*; **Courtney Wrenn**, *Nebulae (emission / absorption)*; **Toby MacLennan**, *Presentation of Playing the Stars*; **Felicity Spear**, *Extending vision: sky-situated knowledge and the artist's eye*; **Vanessa Stanley**, *Surveillance-Surveillance-Surveillance*; **Jim Cogswell**, *Molecular Delirium*.

Contents Vol. 17 no 1 (Spring/Summer 2013)
Clifford J. Cunningham and Günther Oestmann *Classical Deities in Astronomy: The Employment of Verse to Commemorate the Discovery of the Planets Uranus, Ceres, Pallas, Juno, and Vesta*; **Dorian Knight** *A Reinvestigation Into Astronomical Motifs in Eddic Poetry*; **Karen Smyth** *'I specially note their Astronomie, philosophie, and other parts of profound or cunning art': The Use of Cosmos Registers by Chaucer and Others*; **Kirk Little** *Spellbound: The Astrological Imagination of Washington Irving*; **Guiliano Masola and Nicola Reggiani** Σελήνη Τοξότη: *Business and Astrology in the Papyri*; **Reinhard Mussik** *Research Note: Weltall, Erde, Mensch and Marxist Cosmology in East Germany*

Contents Vol. 17 no 2 (Autumn/Winter 2013)
Daniel Brown: *The Experience of Watching:Place Defined by the Trinity of Land-, Sea-, and Skyscape*; **Pamela Armstrong:** *Skyscapes of the Mesolithic/Neolithic Transition in Western England*; **Olwyn Pritchard,** *North as a Sacred Direction? Traces of a Prehistoric North-South Route Across Pembrokeshire*; **Tore Lomsdalen:** *The Islandscape of the Megalithic Temple Structures of Prehistoric Malta*; **Fernando Pimenta,**

Nuno Ribeiro, Anabela Joaquinito, António Félix Rodrigues, Antonieta Costa and Fabio Silva: *Land, Sea and Skyscape:Two Case Studies of Man-made Structures in the Azores Islands*

Contents Vol. 18 no 1 (Spring/Summer 2014)
César Esteban, *Struggling for Interdisciplinarity: Reflections of an Astrophysicist Working in Cultural Astronomy*; **Ronald Hutton,** *Prehistoric British Astronomy: Whatever Happened to the Earth and Sun?*; **Nick Kollerstrom,** *Galileo and the Astrological Prophecy of Manuel Rosales*; **Clive Davenhall,** *Dr Katterfelto and the Prehistory of Astronomical Ballooning*; **Nicholas Campion,** *Celestial Art: An Interview with Geoff MacEwan.*

www.ingramcontent.com/pod-product-compliance
Lightning Source LLC
Chambersburg PA
CBHW060335030426
42336CB00011B/1354